Archaeological 3D GIS

Archaeological 3D GIS provides archaeologists with a guide to explore and understand the unprecedented opportunities for collecting, visualising, and analysing archaeological datasets in three dimensions.

With platforms allowing archaeologists to link, query, and analyse in a virtual, georeferenced space information collected by different specialists, the book highlights how it is possible to re-think aspects of theory and practice which relate to GIS. It explores which questions can be addressed in such a new environment and how they are going to impact the way we interpret the past. By using material from several international case studies such as Pompeii, Çatalhöyük, as well as prehistoric and protohistoric sites in Southern Scandinavia, this book discusses the use of the third dimension in support of archaeological practice.

This book will be essential for researchers and scholars who focus on archaeology and spatial analysis, and is designed and structured to serve as a textbook for GIS and digital archaeology courses.

Nicolò Dell'Unto is an Associate Professor at the Department of Archaeology and Ancient History at Lund University. He is the director of the Lund University Digital Archaeology Laboratory DARKLab. Since August 2019, he is visiting Associate Professor at the Department of Collection Management at the Museum of Cultural History, University of Oslo. Dell'Unto's work focuses on Archaeological Practice, Landscape analysis and 3D visualisation in archaeology.

Giacomo Landeschi is an Associate Professor of Archaeology and works as a researcher in the Department of Archaeology and Ancient History at Lund University and as a research engineer at Lund University Humanities Laboratory. Alongside GIS, his research interests include Sensory Archaeology, Archaeological Method and Theory, Satellite and Aerial Remote Sensing, Landscape and Field Archaeology.

Archaeological 3D GIS

Nicolò Dell'Unto
Giacomo Landeschi

Routledge
Taylor & Francis Group
LONDON AND NEW YORK

Cover image: Cover Archaeological trench recorded in 3D during the field investigation at the archaeological site of Kämpinge and visualized in a 3D GIS. Courtesy of the Kämpinge project.

First published 2022
by Routledge
2 Park Square, Milton Park, Abingdon, Oxon, OX14 4RN

and by Routledge
605 Third Avenue, New York, NY 10158

Routledge is an imprint of the Taylor & Francis Group, an informa business

British Library Cataloguing-in-Publication Data
A catalogue record for this book is available from the British Library

Library of Congress Cataloging-in-Publication Data
A catalog record for this title has been requested

ISBN: 978-0-367-47209-2 (hbk)
ISBN: 978-0-367-47207-8 (pbk)
ISBN: 978-1-003-03413-1 (ebk)

DOI: 10.4324/9781003034131

To Johanna, Livia and Erik (ND)
To Carmen (GL)

Contents

Illustrations

Figures

Tables

Foreword

Since its introduction into archaeology in the late 1980s, GIS technology has become a central tool within many areas of the discipline. Not surprisingly things have changed dramatically since then including the functionality of the technology and the ways in which practitioners integrate both the capabilities and results into their ways of thinking about and describing the past. What hasn't changed, however, is that to do archaeology is to confront the spatial realities of being human, the very essence of our daily lives is to interact with each other and our environments within spatial contexts, constraints, and understandings.

In the early days of archaeological GIS it was clear that these were powerful tools for the management, analysis, and presentation of complex datasets. Cultural Resource Managers quickly picked up on the potential of the technology and established GIS-based monuments records. The other early theme was 'landscape archaeology' which was much more controversial as the post-processual theoreticians criticised the perceived embedded objectivity of GIS-based analysis. The response based on attempts to 'humanise' such digital landscapes through movement and visibility studies made some difference and these still underlie much GIS-based analysis today.

One long recognised drawback of GIS has been the inability to work in a 3D environment, which again in terms of trying to model human experience is a major problem. A 2D (or even 2.5D) landscape can only ever be a partial representation of reality. Often running in parallel to GIS developments in archaeology was the field of Virtual Reality and 3D modelling, albeit often focused on buildings and urban settings. Although practitioners from these two sub-disciplines would come together occasionally it is relatively recently that we see the development of integrated approaches.

The broader descriptor of Spatial Technologies now includes methods and approaches which allow the integration of traditional GIS and 3D modelling, hence the timeliness and importance of this book. For many years the authors have been developing their methods and testing them in a range of situations from excavations through landscapes to buildings and their experience and enthusiasm shines through these pages. The chapters cover all aspects of 3DGIS in detail and with the very useful inclusion of workflow boxes should enable students and professionals alike to come to terms with some complex technical detail. The case studies, including well-known sites such as Çatalhöyük and Pompeii, together with the many and lavish illustrations bring the technicalities to life and show the depth of the authors' knowledge and experience.

While there is always the danger that any book which includes technical detail will eventually become outdated, the fact that this is the first to address 3DGIS will ensure its relevance and importance for many years to come. In their final chapter the authors address future developments and include interesting areas such as the senses, movement and perception, aurality, and acoustics, a list that would move us even closer to modelling being human. If these aims are ever achieved I'm sure it will be these authors amongst those who lead the way into these new 3D recreations of past worlds.

Gary Lock
Oxford, September 2021

Preface

This book is the result of a joint effort made by the authors in an attempt to fill a gap that has long characterised the discipline: the divide between GIS and 3D data visualisation/modelling. For more than 20 years digital archaeologists have been treating Geographical Information Systems and 3D technology as two distinct sub-disciplinary domains, with GIS specialists focusing on tasks such as digital mapping, map analysis, georeferencing archaeological information, creating maps of site distribution, or performing geo-statistics. 3D modellers, on the other hand, have been more concerned with visualisation, rendering, the reconstruction of archaeological artefacts, sites, and landscapes for the purpose of data dissemination, and the reconstruction of archaeological/cultural heritage sites.

In light of recent advances in the production and management of 3D geospatial data, it is now time to reconsider the role of GIS and the contribution that 3D technology can provide to the study of the archaeological record, in a multi-resolution perspective. As has been pointed out several times, reality is three-dimensional (or even four-dimensional if we consider the role played by temporality which characterise the very nature of archaeological research). At the same time, it is crucial for archaeologists to engage in the study of material culture from a contextual perspective. To undertake this endeavour it is therefore necessary to examine the archaeological context as a three-dimensional space where objects and actions took place. It is not just the space of material products of human and environmental activity but also the palimpsest of non-material manifestations of human presence and natural phenomena that are difficult to record and communicate with current data representation standards. The (pre) historic space is characterised by people's movement, encounters, and affectivity, but also by conditions of perception influenced by factors such as seasonality, weather, or illumination. A big challenge for archaeologists nowadays is to rethink the modalities with which they approach the complex reality of the past.

Although this book does not by any means constitute a panacea for all the issues connected to data representation, it at least aims to stimulate readers to reflect on new opportunities for representing and analysing geospatial information by combining multiple sources. With this purpose in mind, the authors will introduce the readers to an integrated approach to GIS and 3D data representation, trying to bridge the gap between two different sub-disciplines and overcome the disciplinary divide that often prevents GIS specialists from constructively dialoguing with 3D modellers and vice versa.

Nicolò Dell'Unto
Giacomo Landeschi
Lund, Sweden

Acknowledgements

This book comes as a result of almost a decade of research activities conducted by the authors at Lund University, in collaboration with several parties including partner institutions, archaeological companies, museums, individual researchers, and specialists from different disciplines.

It would not have been possible to complete this volume without the help and constant support of many colleagues. Therefore, we wish to thank all the colleagues within the Department of Archaeology and Ancient History at Lund University for the constructive feedback on the volume. In particular, Anas Al Khabour, Åsa Berggren, Adam Boethius, Danilo Marco Campanaro, Robert Bergman Carter, Paola Derudas, Domenica Dininno, Fredrik Ekengren, Brendan Foley, Henrik Gerding, Martin Hansson, Jes Wienberg, Peter Jordan, Greer Jarrett, Andreas Larsson, Björn Nilsson, Melda Küçükdemirci, Mikael Larsson, Stella Macheridis, Andreas Svensson, and Anne-Marie Leander Touati.

We would also like to thank the Joint Faculties of Humanities and Theology and the Lund University Library for covering the costs of open access and for supporting this project.

Furthermore, we would like to thank Professor Gary Lock (University of Oxford), Claes Ericson (Interspectral), Helene Wilhelmson (Sydsvensk Arkeologi), Paola Derudas (Lund University), Danilo Marco Campanaro (Lund University), Marcus Söderberg (Lund University), Daniele Ferdani (National Research Council of Italy-ISPC), Marco Cavalieri (University of Louvain), Björn Nilsson (Lund University), Sjoerd van Riel (Norsk Maritimt Museum), Jason Quinlan (Çatalhöyük research project), Marco Callieri (National Research Council of Italy-ISTI), Stefan Lindgren (Lund University Humanities Laboratory), Carolina Larsson (Lund University Humanities Laboratory), Victor Lundström (University of Bergen), Markos Katsianis (University of Patras), Melda Küçükdemirci (Lund University), İlker Işik (Universitesi Edebiyat Fakültesi-Kültür Varlıklarını Koruma ve Onarım Bölümü), Espen Uleberg (HumGis UiO), Martina Polig (Cyprus Institute-STARC-), Håkan Thorén (Statens Historiska Museer) James Taylor (University of York), Anders Follin (LUNARC-Lund University), Roger Groth (GIS Centrum-LU) for the possibility of using their source material and the images included in the volume, and for their invaluable feedback during the writing process. We would like to thank the Lund University Digital Archaeology Laboratory – DARKLab, The Lund University Humanities Laboratory, the HumGis UiO, the Swedish Pompeii Project, the Çatalhöyük Research Project, the 3D Digging Çatalhöyük Project, the Kämpinge Project, the Stora Förvar project, the Uppåkra Project, the Öppenaskar project, Blekinge Museum, the Vång project, the Lund University GIS Centrum, and LUNARC.

We would also like to give a special thanks to the Thora Ohlsson Foundation, the Rausing Foundation, the Swedish National Research Council, the Riksbankens Jubileumfond, The LMK Foundation, and the Crafoord Foundation that generously supported our work and research, making it possible to develop and test most of the techniques and methods described in the book.

We are grateful to the anonymous referees for their valuable comments and suggestions.

Abbreviations

BIM	Building Information Modelling
CAD	Computer-Aided Design
CFD	Computational Fluid Dynamics
CRM	Cultural Resource Management
CT	Computer Tomography
DBMS	Database Management Systems
DEM	Digital Elevation Model
DSM	Digital Surface Models
DTM	Digital Terrain Model
ET	Eye Tracking
GBVS	Game-based visualisation systems
GCP	Ground Control Points
GDBMS	Geodatabase Management System
GIS	Geographical Information Systems
GPR	Ground Penetrating Radar
GPS	Global Positioning System
GUI	Graphical User Interface
H-BIM	Heritage Building Information Modelling
HD	High-Definition
HPC	High-Performance Computing
IDW	Inverse Distance Weighting
IRS	Interactive Reporting System
	Lidar Airborne active sensors
LOS	Line-of-Sight
MAR	Minimal Angle of Resolution
MC	Motion Capture
REVEAL	Reconstruction and Exploratory Visualization: Engineering meets Archaeology
RTK	Real-Time Kinematic
SPP	Swedish Pompeii Project
SRTM	Shuttle Radar Topographic Mission
TIN	Triangulated Irregular Network
UAS	Unmanned Aerial Systems
VR	Virtual Reality

Introduction

Why 3D GIS?

Today, Geographical Information Systems (GIS) represent an indispensable instrument for archaeologists, and it is hard to think of any landscape or field-related project in which a spatial database management system is not employed to handle collected data in some way.

Into the already expansive world of GIS the technologies of 3D modelling have begun to appear and have combined with spatial technologies to expand and diversify the potential of spatial analysis.

In this context, the recent development of new 3D spatial recording technologies, 3D modelling, and the spread of more effective 3D visualisation systems has induced an increasing number of archaeologists to engage with the third dimension and its contextual analysis. The possibility of producing (and using) records capable of describing aspects of the information that were impossible to represent with more traditional documentation methods triggered an experimentation phase in which limits and potentials of these new approaches were assessed. The combination and use of 3D visualisation and GIS had a significant impact on the process of knowledge production and provided researchers with unprecedented opportunities to address new questions and identify novel research trajectories.

How does it relate to archaeology?

Archaeology is, by its very nature, a spatial discipline, and for this reason, it is crucial to rely on effective strategies of data representation to facilitate the interpretative process and to promote a heuristic approach to the study of past human activities. What is the real impact of using 3D data in a georeferenced environment? What is the advantage of connective structured database information to 3D models? Which are the analytical tools currently available in a 3D GIS and how can these be used to improve the interpretation of the archaeological record? This book offers answers to these questions, explores the potentials of these new approaches, and presents multiple examples of how 3D GIS can be utilised to solve a variety of research problems.

Who is this book for?

The main novelty of this work is that it attempts to provide an exhaustive overview of different case studies in which a three-dimensionally informed approach to the investigation of archaeological phenomena can help specialists construct new interpretations

DOI: 10.4324/9781003034131-1

from the archaeological record. Due to an increasing number of practitioners who deal with 3D modelling and 3D data acquisition and widespread use of 3D-oriented tools for capturing archaeological information in situ, it is becoming increasingly important to consider the role of GIS platforms in archaeological practice and their ways of handling spatial datasets. The book engages with theoretical and methodological aspects of 3D visualisation in archaeology and provides practical indications for implementing 3D GIS within archaeological investigations. By using examples drawn from multiple case studies, the text proposes different methods for solving various research problems that are usually encountered when engaging with the study of 3D space. For this reason, readers from a wide variety of professional backgrounds may find inspiration in these pages, including MA and PhD students in Archaeology, specialists in Cultural Heritage, Restoration, and Preservation sciences, and in general any specialists working or interested in engaging with the study of geospatial phenomena. While some boxes (see below, *How to use this book?*) presented in this volume require more advanced knowledge of GIS, others are designed for those who are just starting to engage with 3D GIS.

What will I learn?

This book provides an overview of the different possibilities offered by current 3D GIS platforms in support of archaeological practice. The case studies described in the book are meant to provide examples of how 3D GIS can be used for solving specific research questions or for managing and using archaeological records in 3D space. By spanning from Prehistoric to Medieval sites, Northern Europe to Mediterranean areas, the book illustrates diverse research contexts in which a 3D-informed approach to GIS-based spatial analysis has been employed to address clearly defined research questions. Readers will get an insight into workflows and tool sets that can be reused with different scenarios in mind. On a more practical level, a reader may increase his/her understanding on how to perform several tasks that might include (but not limited to)

- georeferencing a 3D model for performing trench documentation;
- drawing 3D context boundaries and connecting them to the forms in the geodatabase;
- generating and analysing 3D maps of find distribution/density;
- performing advanced visibility analysis in a fully 3D environment.

What is included and what is elsewhere?

As previously mentioned, the book provides a detailed description of research methodologies developed for field activities carried out in Scandinavian and Mediterranean contexts including Italy, Turkey, and Sweden. These methodologies form the basis of a range of case studies in which 3D GIS was used as the main data management system and are meant to provide the reader with practical examples of how 3D can be used in GIS for answering specific research questions. This book should therefore not be read like a technical manual.

Chapter 1 (*Geographical Information Systems in Archaeology*) defines the role of GIS in archaeology and provides a synthesised overview of GIS in archaeology, with a special focus on the role played by spatial analysis and quantitative methods to promote a heuristic approach to the study of the archaeological landscapes and sites. Chapter 2 (*3D Models and Knowledge Production*) focuses on 3D representations and models and it addresses the theoretical and methodological implications connected with the introduction of the third dimension in support of archaeological thinking. By using examples from different case studies, the chapter reviews and discusses the implications of using 3D models for archaeological knowledge-making. Chapter 3 (*3D GIS in Archaeology*) presents different types of 3D GIS platforms and compares those with other forms of 3D visualisation currently employed in archaeology. This chapter provides a brief introduction to 3D spatial analysis and reviews the limits and potentials of different forms of 3D representations that are usually included within 3D GIS. Chapter 4 (*Deploying 3D GIS at the Trowel's Edge*) reviews several examples concerning the use of 3D GIS in support of field investigation and discusses the benefits of engaging with 3D GIS at the trowel's edge, providing indications for producing and managing 3D models in the field. Chapter 5 (*Surface and Subsurface Analysis*) explores the possibilities of 3D GIS for analysing multiple point cloud-derived data in a comparative perspective in order to assess the state of preservation of surface or subsurface layers, especially those affected by the actions of human- or environmental-made threats. Chapter 6 (*Visibility analysis*) discusses the impact of 3D GIS on visibility studies. By using different examples the chapter provides indications for using 3D GIS for the development of experiments focused on visibility analysis. Chapter 7 (*Volumes*) introduces and discusses volumetric visualisation and analysis in archaeology and provides examples and indications on how those can be developed in the frame of 3D spatial analysis. Chapter 8 (*Future developments*) discusses the impact that new techniques and devices currently used for capturing, classifying, and analysing archaeological information are having on archaeological practice.

How to use this book?

The book contains informative boxes designed for presenting some of the methodologies described in the text in additional detail and for providing the reader with practical indications (or possible solutions) for starting to use 3D GIS in support of archaeological practice. Software and methods described in the book represent only a selection of the various possibilities offered by current technology, and advancements in technological development may make the different solutions proposed in the volume somewhat outdated within a short time. However, they are meant to provide the reader with practical examples in which multiple forms of 3D representation are incorporated for solving specific tasks, a practice which we hope will continue into the future, independently of technological advancement. Additionally, many of the case studies include updated descriptions of how their workflow could be improved using current equipment and software.

Chapter 3 contains two informative BOXs which provide solutions for implementing 3D models which lack spatial references into the 3D GIS and introduce practical ways for shifting features in the 3D space. Chapter 4 contains five informative BOXs mainly focused on the implementation and editing of different forms of 3D records within the 3D GIS. Box 4.1 provides indications concerning the use of 3D models for performing micromorphological analyses of archaeological surfaces. Box 4.2 outlines the steps for aligning and importing models representing contexts and features into the 3D GIS that were encountered during the investigation and recorded with other methods and instruments. Box 4.3 introduces the process of setting a reference system for supporting the generation of 3D surface models in the field. Box 4.4 provides indications for importing georeferenced 3D surface models derived from the use of Image Based 3D Models techniques into the geodatabase. Box 4.5 provides indications on how to use 3D shapefiles for outlining contexts and features visible on the 3D surface models, for drawing sections or multicontext maps and for producing volumetric representations of deposits using 3D polygons. Box 4.6 proposes a method for creating and importing 3D GIS boundary (solid) models derived from the interpolation of multiple sets of 3D point clouds or surfaces. Box 4.7 provides a workflow for importing 3D surface models of human and animal remains within a 3D GIS platform. Chapter 5 contains two boxes; the first (Box 5.1) is designed for providing indication on how to assess the issue of surface erosion often associated with archaeological degradation, and the second one (Box 5.2) proposes a methodology for using the 3D models to create thematic maps for supporting the work of conservators. Chapter 6 contains a Box (6.1) that provides indications on how to use the 3D GIS for building vector line-of-sight for connecting observation and target points in the 3D space. Chapter 7 contains two boxes. Box 7.1 presents a pipeline for building a sequence of volumetric extruded layers in which 3D models and legacy data are combined for the interpretation of site stratigraphy, and Box 7.2 provides a solution for creating a 3D grid of vector cubes from an ASCII file for the visualisation of georeferenced volumetric datasets from georadar.

The pipelines described in the boxes can be combined for addressing specific research questions and can be further developed or adapted for supporting the work of specialists operating on site.

1 Geographical information systems in archaeology

1.1 GIS and archaeological theory

If we put together the definitions provided by established and commonly cited authors (Aldenderfer and Maschner 1996; Chapman 2006), Geographical Information Systems (GIS) can be defined as a set of spatial digital tools for managing, manipulating, and analysing information, spatially referenced according to geographical coordinates represented in a Cartesian space defined by x and y (and sometimes z, and even a time variable t) axes. GISs have a longstanding tradition in archaeology, and this subchapter will summarise some of the key points that represented a breakthrough for the discipline from a theoretical and methodological perspective.

An unquestioned point when one thinks about GIS is its impact on archaeology as it has been recalled by Chapman (2006:9). So far, examples of archaeological applications of GIS are countless. The big question that arises is whether GIS can still be considered a method in search of theory (Church et al. 1999) or whether it has developed into something more than a methodology, with its own theoretical culture and infrastructure of discussion; in the words of Gillings and Goodrick (1996), a real place to think. There are several definitions of GIS that have already been formulated and, for the sake of brevity, here we shall simply point out three key aspects that have been highlighted by authors referring to the archaeological application of the toolset. One is the **spatial dimension** of the phenomena that are described and analysed: by its very nature archaeology is a discipline where the material manifestation of the past is examined in its contextual relationship with the surrounding environment. Spatiality thus becomes a pivotal aspect for describing any piece of archaeological information. **Structured information**, typically represented by Database Management Systems (DBMS), is another key element that characterises any GIS platform, and its conceptual framework has long been employed to manage archaeological data of very heterogeneous nature. A third important aspect is **data processing**, defined by the system capability of manipulating existing data to create new informative layers. This "heuristic" component is obviously crucial for an analytic approach to the study of an archaeological context/landscape, and this is an additional reason why GIS had and still has a great impact on the discipline. Interestingly, these three points reflect the concepts of space, representation, and reasoning highlighted in a recent article by Lock and Pouncett (2017), which are the key components of any individual's spatial thinking.

From a theoretical point of view, GIS developed in what can be termed as the "processual" era, a period characterised by a movement known as "New Archaeology" (Binford and Binford 1968; Clarke 1968), with strong emphasis on quantitative and

DOI: 10.4324/9781003034131-2

inferential methods, and inspired by the methodological strands introduced by New Geography (Kohn 1970). These methods started to play an increasingly prominent role in the study of the archaeological record, making several authors (Wescott and Brandon 2003) describe GIS as the most powerful tool to be applied to archaeology since the introduction of radiocarbon dating systems. Again, following Harris and Lock (1990) quoting a British government report, the impact of GIS in archaeology was defined as "the biggest step forward in the handling of geographic information since the invention of the map".

As it has been pointed out several times, spatiality is a significant component of archaeology as a discipline, and the issue of data representation has always been one source of big concern among specialists. Mapping is a very typical example of data representation, and this concept is tightly connected to another keyword of this volume: modelling.

The idea of drawing and placing on a map what has been observed in the field goes far back in time, as archaeologists have always been concerned with formal rules and conventions on how to graphically represent the results of a field survey or excavation (Adkins and Adkins 1989). In this context, GIS provided archaeologists with new opportunities to represent and analyse data collected in the field and dramatically contributed to promote an "explanatory" turn of the discipline, where quantitative approaches and spatial/statistical analysis were employed in support of the study of settlement patterns and site formation.

In this respect, there is a vast literature describing the application of predictive models between the 1980s and the early 1990s (Larralde and Chandler 1982; Kvamme 1983; Pilgram 1983; Brandt et al. 1992; Kvamme 1992) which made GIS gain momentum. Indeed, an increasing number of archaeologists both in academia and in contract archaeology started to use these models for creating maps of archaeological risk (or potential) and defining areas that were more likely to contain archaeological material and so more exposed to the risk of destruction in case of new development. In this respect, predictive models have been traditionally divided into two main categories, focused on quite different scopes: cultural resource management (CRM) and research-oriented or explanatory models.

As a further step, and partially due to the harsh critiques raised by post-processualists, an increasing number of GIS models were developed during the 1990s with the idea of trying to combine the analysis of ecological/environmental factors with the quantitative assessment of human behaviours/cultural elements that had long been discarded in the processual era. A significant example is given by Gaffney and Stancic's work on the island of Hvar, where multiple GIS-based methods were employed to investigate the settlement patterns connected to the chronological phases of habitation in a diachronic perspective. Remarkably, this work constitutes one of the first examples of combined use of predictive modelling techniques together with methods such as least-cost path and viewshed analysis (Gaffney and Stancic 1991).

In this debate, visibility studies and the role of cognition had a prominent part. Several authors within archaeology noted the importance for archaeology to reflect on the centrality of human actors in shaping the past landscape and argued for prioritising the perception and experience of past peoples in any GIS-based analysis. According to several authors (Llobera 1996, Wheatley 2004), one of the main purposes of GIS-based studies should be to foster an understanding of the past landscape through the analysis of this space as a lived-in, meaningful environment where human agency and interaction with the natural ecosystem is a pivotal element. This is one of the reasons why sight, which typically belongs to any human being, was introduced as a possible means by which to investigate the perceptual properties of an archaeological space.

In this context, one of the main limitations has been the inadequacy of conventional data representation schemes (Llobera 2012), where traditional GIS vector primitives (point, line, polygons), and raster datasets are not capable of dealing with the complexity of the landscape under scrutiny. As a consequence, the post-processual discourse has dismissed or at least underplayed the importance and potential of GIS due to the difficulties it has in representing aspects of past realities that this school considered important (Thomas 2000; Tilley 2008). In a recent article (2018) Verhagen stressed the necessity to adopt a different approach, following the proposed distinction between representational and non-representational GIS (Hacıgüzeller 2012). According to this principle, which also draws upon the long-debated problem of developing a more "critical" approach to GIS (Barceló and Pallarés 1996; Pavlovskaya 2018; Perry et al. 2018), one of the main benefits of GIS is to provide archaeologists with multiple models or scenarios that can be built upon the collected data and used for fostering new interpretations with a multivocal approach where research questions are continuously posed to the system and problematised based on a combined use of modelling tools. This contrasts very much with the traditional mapping approach, where a detached, immanent representation of a static space is offered and a reductionist view of the landscape is proposed.

Lock and Pouncett (2017) recently highlighted the importance for archaeologists of focusing more on spatial thinking and reflecting on what real contribution GIS can provide to archaeological reasoning. While GIS undoubtedly provides users with an extraordinary amount of tools for managing spatial information, it is not always the case that users are fully aware of the way selected algorithms and geoprocessing tools actually work. The risk is that archaeological questions are subordinated to the available tools and these, in turn, depend on the GIS software in use in a certain project. This can, in turn, make archaeologists shape their spatial questions based on the available toolkits and not adequately identify the real research priorities.

1.1.1 *Issue of scale*

The problem of the complexity of past realities, and how they struggle to be grasped with the available media, is tightly connected to the issue of scale. Scale deals with the problem of humans' perceptual limitation to connect themselves with phenomena taking place at micro and macro level (Lock and Molyneaux 2006). In regards to archaeological spatial analysis, one common problem relates to the type of representation that is needed to allow archaeologists to grasp the historical/archaeological meaning of a certain space. For example, if the purpose of the analysis is to make a distribution map for all of the classical temples located in a particular region and dated to a specific period, they can be represented as dots on a map. In this case, the vector point primitive is an adequate entity for fulfilling the expectations of a map user who wants to know about quantity and spatial location for the category of classical temples in that region, for instance. In this case, it is the spatio-relational nature of the temple that is under scrutiny, making it possible to investigate their agency in the surrounding space. On the other hand, if one takes the same temple and starts examining the architectural features of which it is composed, it is apparent that such an object is a palimpsest of actions characterised by several temporal phases identified by elements such as columns, plinths, capitals, and so on. The same entity thus becomes a constructed space in which activities took and still take place, a living environment for social encounters. In terms of computer graphics, the vector point primitive will no longer be the right symbol to represent it, and it would be more appropriate to use a 3D boundary model instead to cope with this geometrical complexity. As a consequence, what has been previously analysed as an **object-in-a-space** can be now examined as an **object-as-a-place,** following Tuan's definition of place (1977). The multiscalar nature of today's GIS allows such an approach, in which the same entity is represented either as a dot in a landscape or as a 3D/4D space defined by its own x, y, z, and t coordinates (Landeschi 2019).

1.1.2 High-definition vs. Multi-resolution

In present-day archaeology, one of the big questions is posed by the large amount of information that archaeologists collect from the field, something that has been exacerbated by the introduction of innovative techniques and methods for data acquisition (UASs, image-based 3D Modelling, LIDAR, SONAR data, etc.), more efficient hardware and software solutions (3D GIS, 3D modelling software, game engines, BIMs) and high-performance network/cluster computers (Landeschi et al. 2020). One of the recurring themes in the current agenda is what has been termed as "High-Definition" (HD) archaeology. The term high-definition usually designates digital systems capable of handling data with a very high level of detail. When it comes to archaeological applications (Gowlett 1997), HD was first introduced to define a methodological approach in which the archaeological record is investigated with a significantly high level of detail thanks to the contributions given by analytic methods borrowed from natural sciences. Archaeological excavation is the perfect environment where one can examine sediments and sets of artefacts and ecofacts at micro-scale together with units and stratigraphic layers at middle/macro-scale. The higher the number of intervals in which a single phenomenon can be divided into, the higher its resolution or definition. Such a concept is tightly related to the stances of the 1960s' New Archaeology, when quantitative approaches and statistical methods had a great impact on the discipline. More recently, formal methods of archaeological investigation have been systematically combined so as to broaden the range of material qualities of artefacts and contexts that can be examined. This has led to a new general rethinking of the term "high-definition" archaeology and its proposal as the ultimate goal of important interdisciplinary projects (Raja and Sindbaek 2018).

In this context, the role played by GIS is becoming even more relevant than before. In a posthumanist perspective, where the reality of the past is examined as a continuum where humans, animals and the natural environment are interacting (see Fredengren 2013; Thomas 2015; Braidotti 2016; Díaz-Guardamino and Morgan 2019), it becomes essential to collect, organise, and process data that describe this complex reality in a way that can account for multiple actors and narratives.

An example of this is given by the Çatalhöyük research project, where a GIS geodatabase has been put at the core of the system to manage different types of data including archive reports, older excavation plans, and even field sketches and notes by the excavator during a certain field campaign (Berggren et al. 2015). The combination of data but also experiences and ideas contributed to increasing inclusiveness and multi-vocality among the specialists involved in the field excavation, and this, in turn, reinforced the idea of the archaeological excavation as a place for critical discussion and reflexive approaches (Hodder 1997, 2007).

It is therefore clear that GIS is a pivotal element for archaeological data management, and today's software and hardware performance enable users to explore datasets in a more complete way. In fact, rather than limiting the scope with the concept of HD, it would be more appropriate to consider the multi-resolution nature of the toolkit to fully understand the almost infinite possibilities that GISs offers. There are several types of resolution that can be identified when we analyse GIS performances:

- **Scalar resolution**: this defines the number of scale intervals that the system is capable of handling. Typically, GIS enables users to import and manage data representing very extended portions of land as well as micro-scale artefacts. Multiscalarity is a very unique aspect of geodatabases, where entities of different sizes can be integrated. One critique that has been posed centres on GISs' tendency to focus on the regional scale of mapping and analysis rather than the human scale of the everyday world (Lock and Pouncett 2017). This critique emphasises the difference between an analytical and a phenomenological scale, which characterises our experience as humans. In fact, recent developments in 3D technology make us rethink this aspect and propose novel ways to make GIS overcome the distinction between these two scales of analysis.
- **Temporal resolution:** this defines the number of temporal intervals that can be presented in a GIS project. Archaeological data are by their very nature 4D data, where the temporal dimension is expressed by the number of events visualised by different datasets. One typical example is given by the context unit drawings imported in a geodatabase and featuring a temporal sequence of events that took place in a georeferenced space.
- **Source resolution:** this indicates the number of data formats that can be managed in GIS. Depending on the software solution in use, one system might be more oriented towards raster data, with the possibility of supporting different types of raster file formats. On the other hand, other software is more oriented towards vector data formats, which makes the system capable of managing a higher number of vector data sets. When we talk about source resolution, a further sub/categorisation could be introduced by defining technical specifications associated with certain data formats, that is, raster images, such as radiometric, spectral resolutions, that are the result of the performance of the acquiring sensor.
- **Toolset resolution**: this indicates the number of processing tools available and usable to analyse collected datasets and to possibly produce new data output in the form of new informative layers.

1.2 Data models

Different data standards characterise any GIS. Typically, raster, vector, and table datasets are organised in a structured geodatabase management system and are arranged to fulfil the purpose of a specific project. As a further step in data implementation, topological rules, table relationships, and domains are added to let the users retrieve and examine spatial information in a consistent way.

When it comes to data models, GIS presents different solutions for storing and visualising the geometric and descriptive characteristics of spatial objects. There are two main challenges to be faced by archaeological GISs: to deal with fragmentary information that is often not sufficient to provide the whole picture of the reality that archaeologists are aiming to represent/interpret (see discussion Chapter 8) and to deal with the inadequacy of data representation of most of the current standards. Indeed, starting from the description of current standards is a necessary step before moving forward to the introduction of new paradigms of data representation and modelling. Traditionally, there are three main categorisations employed to describe geographical information collected and visualised in a GDBMS: raster, vector, and table data. The way these categorisation works have been widely explored and a significant number of GIS manuals point out the main features of each one of these data categories (Streatfeild-James 2016 QGIS; Law and Collins 2018 ArcGIS), among them a number of archaeologically focused books (Gillings and Weatley 2002; Chapman 2006; Conolly and Lake 2006). Still, it is important to stress again some of the characteristics of these data categories as this will better introduce the reader to the next chapter's discussion on the combined use of 3D/4D data and their contribution to the development of a more advanced GIS.

1.2.1 Raster data

Raster data consist of a matrix of cells (pixels) organised in a regular grid made of rows and columns. Each raster cell is assigned numeric values that express either levels of intensity associated with a real-world phenomenon in a continuous way, such as temperature, elevation, and so on (https://desktop.arcgis.com/en/arcmap/10.3/manage-data/raster-and-images/what-is-raster-data.htm), or can represent a discrete value used for defining a specific thematic class. Each raster dataset contains a metadata description where different parameters define the characteristics of the data. These include geolocation, spatial resolution, and date of acquisition/creation, giving users a sense of the overall quality of the examined data. For many years raster data have been a key component of most of the GIS-based spatial analysis performed in archaeological contexts (Kvamme 1990; Brandt et al. 1992; Wheatley 1995; van Leusen 1999; Verhagen, P., 2007. Case studies in archaeological predictive modelling (Vol. 14). Leiden University Press.; Herzog 2014), due to the countless possibilities of spatially overlapping and performing algebraic operations between different raster layers and extracting quantitative information that can, in turn, be used to generate new thematic layers. Derivative predictive models, least-cost paths, and viewshed maps are some examples of the new information that can result from the analysis of raster datasets which can all be used in support of archaeological interpretation.

In terms of data representation, some of the current limitations coincide with the intrinsic qualities of the raster dataset itself. These qualities are influenced heavily by the techniques and the devices employed to capture the raw data itself. If one takes an RGB aerial image, for example, its spatial resolution depends on the quality and the performance of the optical sensor employed to generate the dataset. Similarly, elevation raster grids are as accurate as the number of points that have been used to describe the height values of the examined topographic surface.

As a consequence, if one wants to analyse a portion of an archaeological landscape and try to model the topography or the vegetation cover by using raster DEMs and satellite images, they must take into account the unavoidable "gaps" produced as a result of data sampling procedures that characterise any raster dataset. From a theoretical point of view, this aspect contrasts with the idea of representing the archaeological landscape as a continuous space where the supposed emptiness is mapped by integrating different data recording techniques (Campana 2017).

In order to take full advantage of raster information, it is crucial for the user to be aware of the metadata structures and their limitations. Issues such as the spatial scale, the radiometric resolution, and the number of spectral bands are a few examples of parameters determined by the characteristics of the sensor mounted on the acquisition device.

There are also other factors to be considered when it comes to other raster datasets. For example, maps obtained as a result of data post-processing contain human-made choices that can affect the final outcome. Raster DEMs are often derived from different data acquisition techniques including LIDAR and photogrammetry. The raw dataset is then interpolated in order to transform the 3D points into a regular grid of cells where each pixel value is assigned according to a specific interpolation algorithm. Nearest neighbour, Inverse Distance Weighting (IDW), and Kriging are some of the most commonly used algorithms to derive Digital Elevation Models from 3D point clouds (Figure 1.1).

1.2.2 Vector data

Vector data are typically represented by geometrical primitives in the form of points, lines, and polygons.

Points are described by x, y, and sometimes z coordinates that identify the location of the geometrical entity in a Cartesian space according to a predefined coordinate system, either a geographical or a projected one. Each point is typically employed to define a real-world object whose location needs to be accurately specified according to certain coordinate systems in spatial relation with a reference map. A typical example is given by a distribution map of scattered artefacts in plough soil as a result of a field survey, in which each surveyor takes the GPS position for any object collected in the field. The

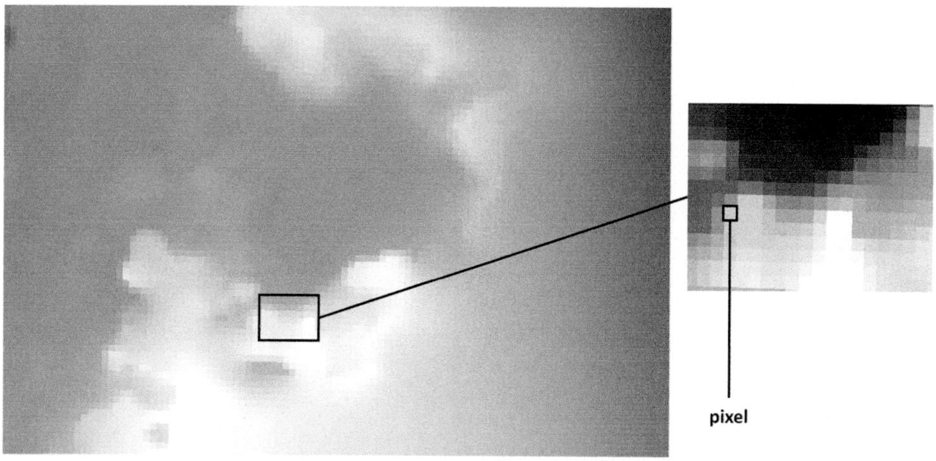

Figure 1.1 Raster dataset representing a Digital Terrain Model. Gradient colour is defined by pixel elevation values of the terrain. The quality of a raster dataset is typically defined by parameters that include its spatial or radiometric resolution.

collected points are then imported in a geodatabase and visualised as georeferenced points in a 2–3D canvas with a satellite map of the study area as a background.

Lines are geometrical primitives that connect at least two points or vertices that are described by x, y, (z) coordinates. Several GIS software packages describe lines and polylines, with the latter used to describe lines formed by multiple segments. Whereas points can be defined as non-dimensional or 0-dimension entities, lines are typically mono-dimensional objects which are useful to represent features such as watercourses and road networks. When it comes to archaeological entities, lines are used, for example, to visually describe on a map the location and the distribution of trenches in a landscape, or to define the orientation and direction of a prehistoric pathway.

Polygons primitives are defined by the combination of lines forming an enclosed space described by two dimensions, length and width. Polygon features are used to represent surface extensions, and their combination allows mapping portions of land which share certain characteristics. Typical examples of polygon features are thematic maps representing land cover information or geomorphology. Archaeological maps are often populated by polygons used to describe different themes, such as site catchment areas, historical regions connected to a city state or land source management (Figure 1.2).

1.2.3 Table data

Table data is an essential component of any GIS project. Spatial information that is gathered during fieldwork is organised so that each single entity is provided with a description that informs the user about certain characteristics of the recorded object, including its geometrical properties and location in the Cartesian space defined by x, y, z coordinates. Attributes are contained in rows or records associated with every single entity that has been created and spatially located in the project database. Tables are often organised in a relational database structure, where data are organised consistently with shared attribute fields used to create relational connections between different tables. This type of data structure makes it easier to query and retrieve information among thousands of stored records. When it comes to the general definition of GIS as a toolset for handling spatial data, it is important to point out that table data represents a category

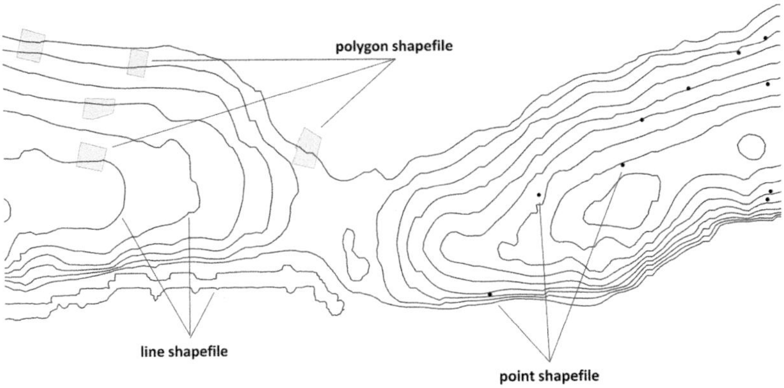

Figure 1.2 Visual configuration of vector data in a Cartesian space, defined by x, y coordinates. GISs typically handle these data as point, (poly)line and polygon shapefiles.

of data models that is not necessarily associated with spatial attributes. There are indeed tables that do not contain any coordinate values but that can still be connected to spatial entities by means of "join" or "relate" operations.

1.2.4 Elevation data (DEMs)

Elevation data constitute another important category of GIS data that provide information about topographic and morphological characteristics of a landscape. There are both raster and vector elevation data that can be managed in a GIS environment. All of these data can be described with the acronym DEM (Digital Elevation Models), indicating datasets representing any type of surface information recorded during data collection. Typically, DEMs include height values belonging to different categories of surface such as ground, vegetation coverage, or built-up areas. Under the umbrella of DEM, there are further categories known as Digital Surface Models (DSM) and Digital Terrain Model (DTM). A DSM describes all the information related to the uppermost surface elements captured by the sensor whereas the DTM results from a filtering operation where only the points belonging to the actual ground surface are left, with the other classes removed. Among elevation data, there are also vector objects associated with height information that can be embedded in the geometry (as an additional coordinate, z) or expressed as an attribute in the table linked to the vector file. These vector files can be, in turn, interpolated in order to generate a Triangulated Irregular Network (TIN), a vector mesh made by triangles derived from the vertices belonging to the origin files (points, lines). A typical way to derive a DEM from traditional printed maps was to digitise elevation spots and contours, respectively, as vector points and lines and then to generate a TIN from the interpolation of both.

An important aspect regarding DEMs, DSMs, and DTMs is that all of these raster files consist of a grid of cells in which the height value is expressed as an attribute (preventing each x, y terrain unit from being described by more than one Z value), and therefore they can be considered as not fully 3D models but 2.5D. This is an important remark that will be further discussed in the following chapters and that represents a crucial difference between traditional and proper 3D GIS, where objects defined by independent x, y, z coordinates are visualised and used in data analysis.

1.2.5 Topology

Another typical feature of many GIS products is the handling of topologies. Following the definition provided by the QGIS guide (https://docs.qgis.org/2.8/en/docs/gentle_gis_introduction/topology.html#topology-errors), topology expresses the spatial relationships between connecting or adjacent vector features (points, polylines and polygons) in a GIS. When topological rules are established, vector objects must fulfil certain conditions in order to maintain a consistent representation of the spatial distribution and relationships among the described entities. A typical example of a topological rule to be fulfilled is the polygon overlap, according to which polygons are not supposed to overlap each other, with respective vertices not included within the area enclosed by the adjacent polygon. An essential element in order to set up topological relationships is the node, which indicates a contact point between different entities. If one takes a road network for instance, a node will indicate the precise point where two (or more) roads meet.

1.3 Current limitations

This concluding part of Chapter 1 tackles current GIS approaches by highlighting issues which prevent these systems from providing a dynamic and coherent means to deal with the complexity of the past(s). The idea is to provide the reader with food for thought by pointing out both material and immaterial aspects of reality that are pretty much absent from current GIS-based research. This discussion will introduce novel methods of data representation and analysis that will, in turn, be discussed in the following chapters. The resulting paradigms will provide a basis for envisioning new ontologies that can be included in a collaborative environment where GIS acts as a pivotal element to collect, visualise, compare, and analyse multiple sources and dynamic processes.

No matter whether GIS is more oriented towards raster or vector data, there are issues that still affect and pose serious limitations to the full understanding of past world realities.

Considering the landscape as an entity characterised not only by natural and anthropic elements but also by actions and activities that took place in a multi-temporal perspective (Ingold 1993; Llobera 1996), we must revise current standards of representation and reflect on which elements are missing from the "big picture". To grasp the complexity of what we call reality, it is essential to take into account all material and perceptual elements that could have been "experienced" by people in the past. This means considering any material object in its three dimensions but even more, all the subtle elements of reality such as fluids, natural, and artificial light that are often neglected in traditional scholarship but that can actually inform us on very particular aspects of the daily life of a medieval village or a prehistoric cave, say. If one takes into account the human sensorium, the possibilities of extending this analysis to multiple sources are manifold: indeed, acoustic, olfactive, and tactile sensations are part of the way one can experience a physical space and can assign a special meaning to it (Landeschi and Betts, in press).

Affordances are another important element that has been recently pointed out as the optimal focus of GIS-based landscape analysis. According to Gillings (2012), the way to integrate affordance into the GIS-based discourse is to reconsider this concept and think about what he termed "experiential affordances". The observation of the relationships between human and environment can better allow to detect "areas of suitability" that can be defined as the space in which certain actions are performed (burial mounds or any other type of visually engaging monument) based on the relation between a certain element of the landscape and the human as a perceiving being. It is clear that the agency of humans in the landscape should be the main target of this kind of analysis. What is less clear is the way the data-capturing process should take place, given the significant limitations for acquiring very diverse forms of phenomena.

To fully conceive the role of GIS in this process, it is important to bear in mind the concept of the model. As recalled by Lock (1995), the very concept of the model promotes the idea of simplifying reality in order to make certain phenomena easier to understand. Indeed, the ultimate aim for a heuristic approach to archaeological data modelling should be not to imitate but rather to "simulate" the past (Forte 2011). This means presenting and connecting those elements that allow archaeologists to experimentally create a possible scenario that could help us to understand how facts of the past took place along with their social meaning. In this respect, the possibility of gaining new knowledge about the past depends upon the quality and the quantity of information that is used to create this simulation scenario. In the light of the very diverse nature of the phenomena that constitute a reality or a context, it is clear that there remains much to be done before we can fill in the current gaps in data representation. Especially if we consider affordances, it is extremely difficult to define which kind of action, feeling, or stimuli can be determined by the landscape in different agents, humans, and animals.

It is therefore crucial to reflect on the current limitations of traditional GIS data models before moving on and introducing new paradigms of data representation. In this respect, there are two main categories of issues: **geometrical** and **procedural**.

Geometrical issues: intrinsic spatial characteristics of the traditional GIS feature's geometry are limiting the possibility of representing real-world phenomena. Typical archaeological problems, such as the representation of a stratigraphic sequence, or subsurface data collected through geophysical prospection, require scientists to consider the issue of volumetric data (see Chapters 3 and 7). Except for a few attempts made in order to manage volumetric information in the context of stratigraphic excavation (Barceló et al. 2003; Bezzi et al. 2006; Lieberwirth 2008), GISs have never been used to systematically handle and analyse relationships between volumetric layers and classes of artefacts/ecofacts contained therein in a quantitative way. Although standards to describe volume in computer graphics exist at least from the late 1980s/early 1990s (Drebin et al. 1988; Kaufman et al. 1993), the procedure for handling these formats in most of the current GIS software is far from straightforward, as will be illustrated in Chapter 3. Another interesting field of application for volumetric data is geophysical prospection, which constitutes a vital and non-destructive investigation method for subsurface contexts. To date, most geophysically derived datasets (typically GPRS data) are imported and visualised in GIS just as raster layers representing surface discontinuities corresponding to different depths from the reference ground surface. In fact, as will be illustrated in Chapter 3 and 7, the potential of these data lies in the possibility of gaining information about the volume of subsurface layers without having to actually excavate them, thereby defining the spatial characteristics of anthropic and environmental actions belonging to specific deposits.

- (**fluids**) Another geometrical issue relates to what is known as computational fluid dynamics (CFD). GIS software is not supportive of any particle-based algorithms or equations that are typically used in computer graphics to simulate fluid movement. This prevents archaeologists from using powerful representational tools that would inform them about submerged sites, perform flood simulation and make quantitative assessments of erosion in threatened underwater sites, just to name a few examples. As an additional problem, CFD could be used to simulate smoke or gases movement and thus represent specific weather conditions at different times, which would be extremely useful for picturing changes in the perception of a certain landscape in a multi-temporal perspective.
- (**light**) On a similar note, natural and artificial illumination affects human movement and trajectories in a landscape in multiple ways. When it comes to perception and sight, it is crucial to reconsider traditional approaches by integrating the effects played by lighting. Although the theme of light in archaeology has been recently treated (Papadopoulos and Moyes 2017), very little attention has been placed so far on integrating light effects into GIS-based spatial analyses, with most of the examples of light simulation being performed with dedicated software in non-georeferenced environments.

- **Procedural issues** relate to data acquisition in terms of hardware/software performances and user-made choices. In this respect, device limitations still constitute an important bottleneck. Despite the unprecedented advances in terms of performance, many computing stations still struggle to handle huge quantities of data such as those collected by high-performance sensors. The Swedish cadastral agency (Lantmäteriet 2021), for example, now allows users to visualise and download multispectral aerial images in raster format with a spatial resolution in which each pixel represents a portion of land of 0.15 m. Whereas this represents an incredible qualitative advantage, opening up for the chance to detect ground anomalies on a lower scale (and thus overcome the typical limitations of traditional remote sensing approaches), new problems arise in terms of computational performance. Even a PC with gaming-level specifications (3.5 GHz CPU, 8/16 GB RAM, 8 Gb graphics) would hardly manage to properly visualise more than four separate aerial images with the previously described resolution.

To cope with this, a possible solution is to take advantage of High-Performance Computing (HPC). However, this requires the establishment of collaborations with specialised centres which have the necessary computational resources. Even though this approach is becoming more widespread, with several archaeological departments starting collaborations with HPC-oriented research centres (White 2016; Dodd 2018; Landeschi et al. 2020; Zheng et al. 2020), it is not clear yet what the implications of handling "big data" would be for future GIS software.

As an additional set of procedural issues to be considered, we must mention all the human-made choices that lie behind every single data acquisition/recording process. Even though many of the intrinsic qualities of the dataset depend upon the sensors or the technical specifications of the acquisition device, there are still many choices made by the surveyor which affect the final product. This is an important aspect that is worth exploring, especially in a time when software ease-of-use makes a larger set of users approach data acquisition, with the very (naively) optimistic perspective of getting an accurate 3D model of an archaeological feature in just a few steps with minimal effort. One of the biggest downsides brought about by this sort of "democratization" of archaeological field documentation is the risk of forgetting the human factor in the construction of what we could term "wise" models.

As an example, in image-based 3D Modelling approaches it is not just the digital camera and the software that guarantee an accurate 3D model of an object that needs to be documented, but also parameters that depend on the position and movement of the user in the field. Distance, field of view, and image overlap are examples of choices that need to be made by the user during all the phases of photographic data acquisition, and it is therefore crucial for the surveyor to be aware of the consequences each single choice will have on the data post-processing.

Similarly, the parameters set to align a point cloud, to generate a mesh and finally to export a textured 3D boundary model dramatically affect the possibility to correctly import and visualise the documented object in GIS. This may result in a 3D boundary model with a lot of geometric redundancy, preventing the GIS software from correctly importing and displaying the dataset. It is therefore important for the user to be aware of the right parameters to be set in order to produce a 3D model suitable for the geodatabase handling and visualisation.

1.4 Summary and conclusions

This chapter summarised the state of the art in GIS in archaeology. Starting with an examination of methods and techniques for the acquisition and the processing of geo-spatial data, the chapter made a review of the main theoretical strands that influenced the discussion and the application of GIS-based methods over the years. Most of the concepts and terms introduced here do not constitute a novelty per se, but their description is an essential step for clarifying current limitations and biases related to the representation of data in most of the currently available GIS software packages. Scale, resolution, and data formats are concepts that need to be carefully considered when planning a GIS-based project. These concepts are also relevant for introducing the use of 3D models and three dimensionality in the representation of archaeological data.

Indeed, current data representation (vector primitives and raster grids) in GIS have some limitations, and authors have recently emphasised the inadequacy of most of the software packages for dealing with the complexity of the archaeological space (Llobera 2012). It is about time that standards be revised and rediscussed based on the introduction of geometrically complex entities such as 3D surface, boundary, and volumetric models as a result of more advanced field acquisition techniques.

2 3D models and knowledge production

2.1 3D digital modelling in archaeology

In computer graphics, 3D visualisation is a method that employs 3D contents (or 2/2.5D or textual contents) displayed in a Cartesian 3D space (defined by x, y, z coordinates) for extracting meaningful information, communicating a message, clarifying spatial or topological relations or for better illustrating phenomena or processes, through the use of interactive graphics and imaging (Kaufman et al. 1993).

3D modelling is the process of creating a geometrical representation of an object in 3D (Remondino and El-Hakim, 2006). The process of 3D digital modelling can be developed using (or combining) multiple types of techniques, which lead to the generation of 3D digital objects that can be used in multiple ways. 3D digital models are subject to several classifications (Foley et al. 1990) and can be categorised according to different functions, affordances or uses. 3D models should always be considered as the result of an interpretation process, whose characteristics reflect the intent of the interpreter to highlight specific aspects of the material being represented (Campana 2014:8; Dell'Unto 2018) (Figure 2.1). To use Nordblad's words, a 3D model can be defined as a "*summary and simplification of a physical thing, which helps to explain something complicated in one view*" (Nordblad 2012:245).

In archaeology, 3D digital models permit experimenting with new forms of intellectual engagement, increasing the possibilities of identifying information and revealing new evidence. The most widespread types of 3D models currently in use in archaeology can be divided into **surface**, **boundary,** and **volume** representations (e.g. voxel and spatial decomposition models). These types of 3D models can be utilised in support of archaeological practice, and in the last decade, their use has increased exponentially, reaching almost any area of the discipline.

Surface representations are 3D objects constituted by points, lines and surfaces, and represent the shell of the object rather than its internal characteristics (Lock 2003:152; Hughes et al. 2014). Despite the limits of these type of models in providing only a partial description of the object being represented, their recent spread triggered an intense phase of experimentation which led to the definition of new archaeological practices (Callieri et al. 2011; Forte et al. 2012; Opitz and Nowlin 2012; Taylor et al. 2018), moving archaeology more and more towards a collaborative and immersive discipline (Berggren et al. 2015). **Boundary representations** are 3D models characterised by a topologically consistent surface. These types of models function as solids rather than shells (de Cambray 1993:342), and allow extracting information such as volume, mass, and centre of gravity (Lock 2003:152). Boundary models are often used in archaeology

DOI: 10.4324/9781003034131-3

to create virtual interpretations of artefacts and monuments or for reviewing the stratigraphic relations between the deposits encountered during the investigation. **Volume representations** such as **voxel** and **spatial decomposition** models have the capacity to describe the object's inner space. Those models are particularly suited to represent geological data (de Cambray 1993:343), and their use is very significant for representing data sets acquired using instruments that allow recording and assigning different values to the materials which compose the object.

3D models increase our ability to review the formation process and to observe how spatial data relate to each other across time and space (Figure 2.1). Their spread has important theoretical implications affecting the way archaeologists record, analyse and interpret the past. As pointed out by Campana, their implementation should not be considered the end but rather the means for achieving a deeper understanding of the materials encountered during the investigation (Campana 2014:10). 3D models are used to support researchers and scholars across the entire archaeological sector and their spread has a substantial impact on how we perceive and connect information coming from different investigation methods.

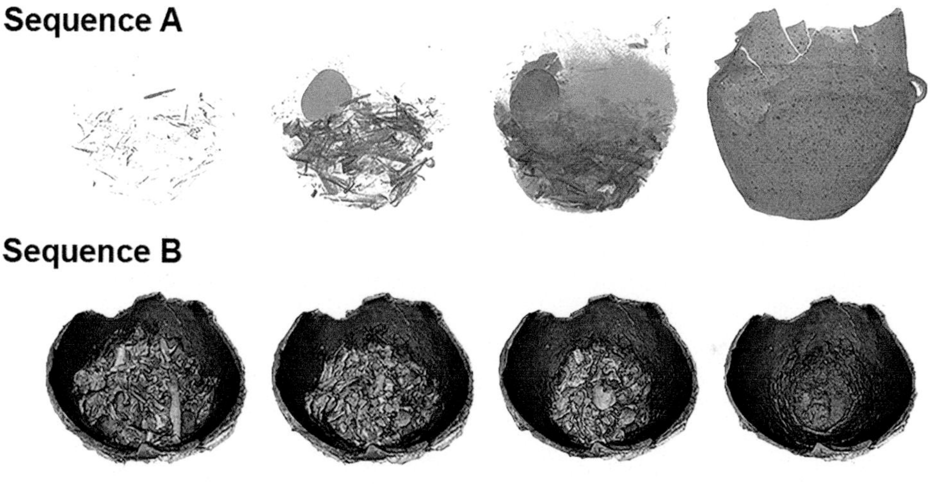

Figure 2.1 3D volumes and 3D models of Bronze Age urn burials from Domsten, Sweden. The urns were scanned using a medical CT scanner (A sequence) and image-based 3D Modelling techniques (B sequence). A sequence shows (from left to right) (I) only the highest density portions of the bones within the urn as well as the bronze awl in red (as it has a very high density) (2) medium density material is visible, (3) very-low-density material (sand and very porous bones) is visible in lighter green colour, (4) the urn is separated from its context in the virtual environment and viewed at medium density level revealing minerals within the clay as red dots. B sequence shows a selection of the layers documented during the stratigraphical excavation within the urn. Each new layer of bones and objects were documented using Image Based 3D Modelling, for a total of 17 sequences. All the separate layers have some overlap with key bones identifiable in more than one sequence to facilitate integrating them with the CT volumes in a common environment (3DHOP). Image courtesy Claes Ericson (Interspectral), Helene Wilhelmson (Sydsvensk Arkeologi) and Paola Derudas (Lund University). CT scan Acquisition: Marcus Söderberg (Lund University), 3D data processing and volumetric visualisation: Claes Ericson (Interspectral) 3D data acquisition and Image Based 3D processing: Paola Derudas (Lund University).

2.2 Background and context

3D models (in physical form) have been employed in archaeology since the 18th century. The use of plaster copies and wooden models was a widespread practice in museums collections and academic environments (Huvila 2006). These objects were (and still are) used for teaching, for public display, for research, and were efficiently employed for raising interest in the study of antiquities as well as for supporting the production of new knowledge (Dell'Unto 2018). Since their introduction in archaeology, 3D models have been distinguished into two main categories: *interpretative and reality-based models.* The former refers to hypothetical reconstructions of how a specific object (e.g. a monument, an artefact or a site) might have looked like in the past. The latter instead refers to 3D representations which display the object as it is at the moment of data acquisition (Figure 2.2).

Both types of models were the result of detailed calculations and were often accompanied by accurate reports which described the process leading up to their creation. An interesting example of such a practice can be found in Giovanni Altieri's temple of Isis, commissioned by the King of Sweden after his visit to Pompeii (1782). The model was the result of a meticulous use of archaeological documentation, and its creation illustrates the creator's desire to represent not just one monument but a building in spatial relation with its surroundings (Kockel 2004:144). These models were the product of

Figure 2.2 Examples of (digital) interpretative (1) and (digital) realistic model (2) of the house of Caecilius Iucundus, Insula V, 1 Pompeii. On the left side of the image, the 3D surface model generated after interpolating the laser scanning data. Once acquired, the points clouds were processed in Meshlab and published on the web using 3DHOP. On the right side of the image, a 3D interpretation of the house of Caecilius Iucundus generated using Blender and rendered in Unity 3D (Dell'Unto et al. 2015).

intense intellectual work, which often involved multiple experts and took into account different aspects of the monument's documentation.

Physical models were produced to represent interpretations, to reproduce the original characteristics of objects, monuments, or sites, and to portray social space (Nordblad 2012). Several of these models contain information that is no longer available in the originals, and for this reason, they have often been used (or re-used) as a primary source of information.

An interesting example of this process can be found in the work of Felice Padiglione, who at the end of the 19th century was commissioned by Giuseppe Fiorelli to create a model of the city of Pompeii. This work is currently exhibited at the Museo Nazionale in Naples and despite the small dimensions (scale 1:100), it is considered an important archaeological reference which "*has not yet, by any means, been exhausted*" (Kockel 2004:149), and which is often used (physically and digitally) for verifying hypotheses and constructing new interpretations (Figure 2.3).

Despite their physical limitations, these models were the result of intense technical and intellectual work, involving multiple stakeholders and including archaeological and historical documentation. The impact and ongoing importance of these productions are connected to the quality and transparency of their executions. We can rely on these models because in most cases (1) we can reconstruct the history of their creation, (2) we can retrieve information about the context in which they were made, and (3) we can identify the ways in which the model can be used and what it can offer to researchers.

Physical models were well integrated within the archaeological process, and their creators were deeply familiar with the target site or artefacts, and often skilled heritage practitioners, capable of reading, understanding and transforming different types of archaeological and architectural records into 3D models.

Figure 2.3 Model of the city of Pompeii made by Felice Padigioni and currently exhibited at the archaeological museum in Naples. The model is an accurate reproduction in scale 1:100 of the site during the 19th century. Image courtesy of the Italian Minister of Culture, National Archaeological Museum of Naples, and Danilo Marco Campanaro. Su concessione del Ministrero della Cultura-Museo Archeologico Nazionale di Napoli.

2.3 3D models in the digital revolution

In the last two decades, the development and diversification of computational resources and applications has prompted the development of methods for digitally manipulating and visualising 3D contents.

In the early 90s, archaeologists started considering these instruments as tools for increasing the perception of archaeological material and for supporting new forms of interpretation and practice. Paul Reilly introduced the term Virtual Archaeology, referring to the use of 3D modelling to simulate the archaeological formation process (Reilly 1991:133). The possibility of using computers to construct and visualise archaeological interpretations was perceived as an alternative and dynamic solution which provided archaeologists with more accurate observations to draw upon when formulating questions, as they could now rely on a virtual representation of the relationship/s between spatial phenomena as a visual reference for their enquiries (Renfrew 1997:7).

Although the opportunities offered by 3D modelling for supporting the interpretation process were significant, at the beginning of 2000 most of the papers discussing Virtual Archaeology focused mainly on computer graphics innovations and public engagement, with very limited impact on archaeological theory and methods (Hermon 2008:37).

One of the issues that prevented the spread of 3D models was the lack of transparency in producing these models. To be used in support of archaeological research, they needed to include information about how they were created and for what purpose. After a period of experimentation, several methodologies for encouraging best practice were developed. Of these, perhaps the most significant was the London Charter for computer-based visualisation of cultural heritage, published in 2006 to encourage methodological rigour (http://www.londoncharter.org). Subsequently, in 2011 the Seville Principles of Virtual Archaeology were developed for increasing the conditions of applicability of the London Charter in archaeology (http://sevilleprinciples.com/). The establishment of such guidelines led to the creation of specific formal languages for tracking the virtual interpretation process undertaken during the 3D reconstruction (Niccolucci and Hermon 2010). One of the latest examples of this phenomenon is the extended matrix; a formal language that focuses on creating a strong link between archaeological documentation and virtual interpretation (Demetrescu 2015). Since the establishment of this approach, several experiments have been developed for testing the potential of computer-based visualisation as a tool for supporting the interpretation and modelling of archaeological data. One of the latest experiments was conducted as part of the international project "*VII Regio. The Elsa Valley during the Roman Age and Late Antiquity*", an Italian-Belgian research project coordinated by UCLouvain and developed in collaboration with the National Research Council of Italy-CNR ISPC. The project employed different 3D acquisition and visualisation technologies to test the interpretive potential of 3D modelling structures and artefacts retrieved during the investigation (Ferdani et al. 2019) (Figure 2.4).

A remarkable example of the use of 3D models in support of archaeological thinking is the work of Frischer and Fillwalk, whose 3D reconstructions and computer simulation systems were combined and used for investigating the relationship of different monuments in the Campus Martius in Rome. The main goal of this work was to use a heuristic approach for evaluating the validity of the Buchner thesis, which claimed that the obelisk and Ara Pacis were intentionally connected to have a solar alignment on

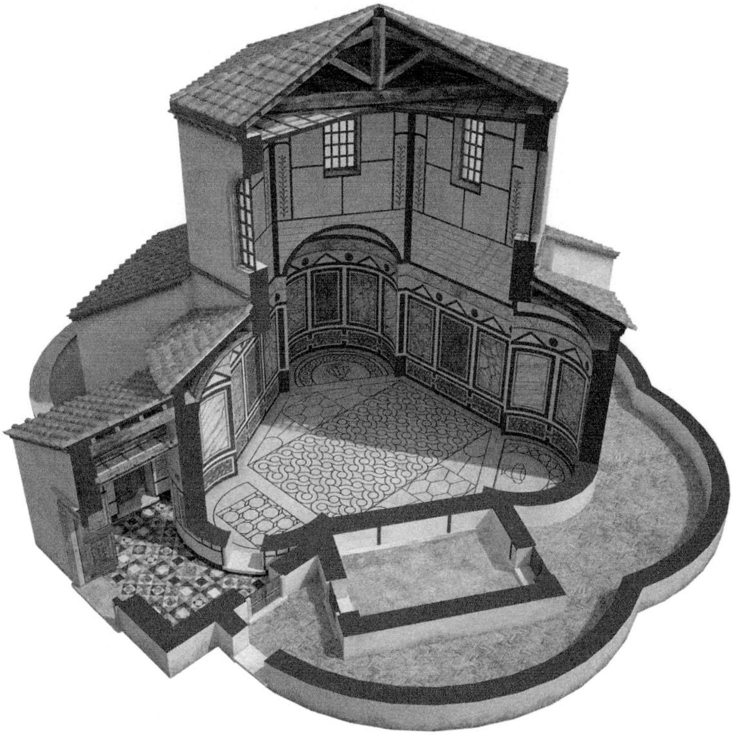

Figure 2.4 3D interpretation of the Roman Villa of Aiano, San Gimignano-Siena. 3D interpretation and reconstruction developed by the University of Louvain in collaboration with the National Research Council of Italy-CNR ISPC. Image courtesy of Daniele Ferdani and Marco Cavalieri.

Augustus' birthday (Frischer and Fillwalk 2013:341). The development of this method allowed them to perform accurate observations which clarified the structural and symbolic relationship between the monuments at the time of their construction (Frischer and Fillwalk 2013:344), demonstrating the advantages of using 3D models for simulating a range of possible scenarios. This work represents an excellent example of the use of virtual environments for defining (rather than presenting) an archaeological interpretation.

In the volume Cyber Archaeology, Forte describes Virtual Archaeology as a static and graphic oriented discipline, incapable of providing archaeologists with the tools for simulating a potential past (Forte 2010:10). By introducing Cyber Archaeology Forte shifts the focus on simulation rather than visualisation, describing a discipline where interpretations occur as a result of a dynamic exchange of information between the user and the digital environment (Forte 2010). As pointed out by Lanjouw, Cyber Archaeology has many similarities with Reilly's original description of Virtual Archaeology. However, Forte's work is strongly rooted in Bateson's theory on learning (Lanjouw 2016:7), his experiments going far beyond the simple visualisation/interaction, and he relies on the assumption that the migrations of data within a cyber domain creates more advanced knowledge and produce unexpected results (Forte 2010:9).

Thomas Levy defines Cyber Archaeology in a broader sense, referring to an interdisciplinary field that results from the joint effort of computer scientists, engineers and digital archaeologists (Levy et al. 2012; Levy and Jones 2017:2). By referring to its capacity to include multiple representations of the archaeological investigation process, Zubrow describes this new field as the link between "scientific and interpretational archaeology" (Zubrow 2010:2).

The development of this new approach was triggered by the spread of new technologies for 3D recording and visualisation. Spatial technologies such as image-based 3D Modelling, laser scanning techniques, global navigation satellite systems, Unmanned Aerial Systems (UAS) allowed documenting in high resolution and with great accuracy almost any type of archaeological data. Differently from computer graphics programs such as Blender or Autodesk 3DSMax, which are mostly employed in archaeology for modelling 3D interpretations of sites, monuments, or artefacts, 3D spatial recording techniques primarily generate high-density point clouds, which can be further processed for creating different types of 3D representations (Figure 2.5).

These new acquisition methods deeply impacted archaeology, and their spread reduced the distance between the physical and digital space (Dell'Unto 2016), with a substantial effect on several areas of archaeological activity (Opitz and Limp 2015). Edgeworth underlines how the use of computers for modelling and visualising archaeological data impacts the discipline's organisational structure (Edgeworth 2014:55). He also argues that computer representations (mainly referring to aerial photos and satellite images) can be used for challenging theories and ideas at the same level as tangible materials, reinforcing the idea that these representations have an impact beyond the process of discovery (Edgeworth 2014:50–55). In addressing the impact that digital technologies and 3D recording are having on archaeological practice, several authors have discussed the importance of generating data as a result of a direct and non-mediated engagement with the material being recorded (e.g. Dell'Unto 2016:319; Powlesland 2016).

The act of recording archaeological information is considered by many archaeologists a part of the process of understanding it (Giuliani 2008). Therefore, a major risk in distancing the virtual and the physical could be a loss of intellectual engagement with the materials encountered during the investigation. This reinforces the idea that 3D models and 3D visualisations should not be regarded as an alternative practice but as a fundamental "new" part of the way archaeologists investigate the past.

The spread of new investigation techniques based on scientific methods and advanced data recording prompted the creation of large datasets of information that are very difficult to manage with traditional data management systems. The use of 3D models as

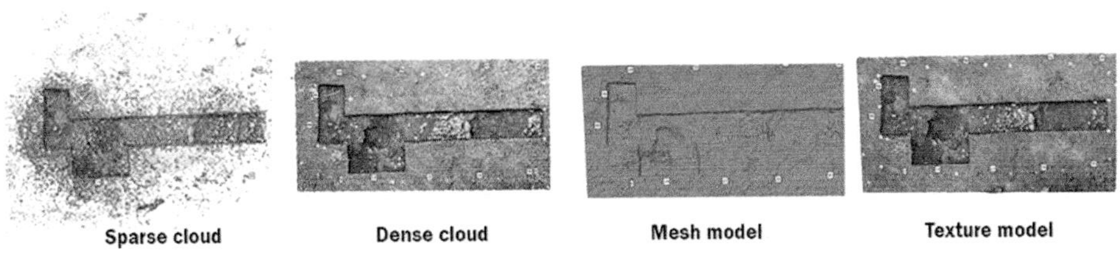

Figure 2.5 The image illustrates the process of generating a 3D model using Image Based 3D Modelling techniques and different outputs.

spatial media for representing different records collected during the investigation represents a key element in current archaeological practice (see Figures 3.1, 4.2, 4.5, 7.3, and 7.5).

2.4 Different models, different affordances

2.4.1 3D surface and boundary models

As previously discussed in Section 2.1, 3D surface and boundary models are among the most common outputs from the use of 3D modelling and 3D digitisation techniques. Virtual interpretations are often created using boundary representations and surface models are frequently employed for representing context interfaces. Dense points clouds and bi-dimensional images (resulting from the employment of active and passive sensors such as digital cameras or laser scanning) are frequently transformed into 3D textured models to be visualised within different platforms (such as 3D GIS or game engines) for achieving various forms of representations.

Despite this range of acquisition strategies, their creation often follows two specific phases: data gathering and data modelling. Each of these steps must be carefully planned and involves several actions and choices made by one or multiple operators. Despite the different guidelines available through several cultural heritage institutions, the establishment of a universal approach for creating 3D surface or boundary models in support of archaeological research is very complex. Each 3D representation demands the definition of a specific methodology that should reflect the research questions posed during the investigation. The process chosen for creating the 3D models will inevitably affect their final output, assisting (or hindering) scholars in the development of their research.

As a (very) general indication, it is important to (1) perform an accurate analysis of the object to be modelled or acquired, (2) examine the features which characterise the object (e.g. geometry, materials, colours), (3) establish (before acquisition or modelling) how the model will be employed once created, (4) analyse all the available sources to (5) identify the right modelling techniques and/or acquisition instruments to employ, and (6) adopt a specific formal language for tracking the interpretation process used during the virtual interpretation.

These steps should lead to the generation of models which are often employed:

- As a background for producing digital drawings.
- As the primary source of documentation.
- For achieving representations of monuments and complex architecture (e.g. visualisation of the external and internal parts of a building or/and underground environments).
- For analysing the shape and morphology of archaeological artefacts and materials.
- For building virtual interpretations (or reconstructions) of sites and monuments.
- For reviewing the stratigraphic relation among different field deposits.

The way these models are constructed inevitably affects their characteristics and appearance, and for this reason, their creation should always be undertaken (or supervised) by a cultural heritage specialist. These types of models have proven to be excellent records in support of disciplines such as forensic and bio-archaeology (Wilhelmson and Dell'Unto 2015; Seguchi and Dudzik 2019), artefact analysis (Scalas et al. 2018) and landscape, field

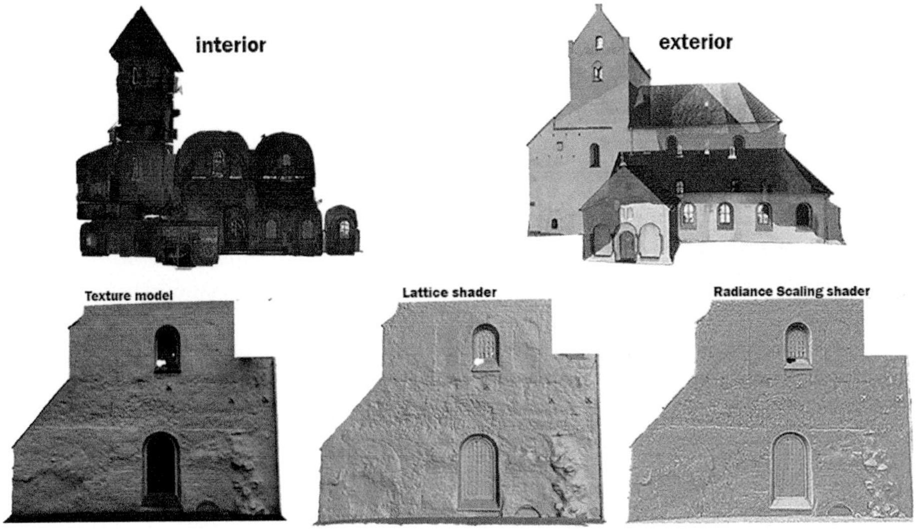

Figure 2.6 3D surface models of part of the structures of Dalby church, Sweden. In the upper part of the image, the church's interior and exterior was acquired using Laser scanning techniques. Below, the 3D model of the west wall generated using Image Based 3D Modelling techniques, visualised in Meshlab using different shaders. Data acquisition and data processing: Stefan Lindgren (Lund University Digital Humanities Laboratory) and Nicolò Dell'Unto (Lund University).

and building archaeology (Callieri et al. 2011; Opitz and Nowlin 2012; De Reu et al. 2013; Merlo 2016; Polig 2017; Katsianis et al. 2021). 3D surface and boundary models have been used for expanding our perception of the physical space (Forte et al. 2012) as well as for increasing our understanding of archaeological materials (Horn et al. 2019) (Figure 2.6).

2.4.2 Volume representations

3D volume representations such as voxel and spatial decomposition models result from a more sophisticated modelling process, and their creation is often connected with particular visualisation needs. These representations can be generated by interpolating virtual interfaces of archaeological contexts (for more examples see Chapter 3, 4 and 7) or by using recording instruments that identify and document the different materials contained within the enclosed space of an object or an archaeological deposit (e.g. computerised tomography scanners and geophysics). Despite the limited use of volumes in archaeology today, the formulation of new research questions, the spread of new acquisition instruments and the development of new analytical tools has triggered an experimentation phase which is providing results beyond simple data visualisation (Minozzi et al. 2010; Huisman et al. 2014).

Among the different techniques and sensors for producing 3D volumes in archaeology (most of these are discussed in detail in the upcoming chapters) Computer Tomography (CT) scanners have recently seen a particularly interesting development in archaeology.

Since their introduction in the 70s, X-ray CT scanners have always been used in support of archaeological investigation. The development of x-ray CT allows researchers to virtually dissect archaeological materials, providing them with the opportunity to perform more accurate analysis (Hughes 2011). If initially this technique was mainly relegated to the study of mummified remains, today its use has been extended to different types of objects and materials.

CT scanning techniques have been used to identify text on clay tablets (Applbaum and Applbaum 2005), to study bronze artefacts (Rossi 1999), for inspecting cremation urns (prior to physical examination) (Minozzi et al. 2010) and more recently, Micro CT scan techniques were employed in ceramic studies for detecting clay fabric microstructures that are indicative of specific craft practices (Kozatsas et al. 2018). In the field, X-ray tomography has been employed to identify the shape and position of archaeological materials contained inside soil blocks extracted during the excavation. These provided archaeologists with important indications (before excavation) concerning spatial position, shape and state of preservation of artefacts and materials (Re et al. 2015) (Figure 2.7).

Volume visualisations created with these techniques allow archaeologists to identify and observe patterns and relationships that are often too subtle or scattered for us to perceive by other means. By visually contextualising the results of these observations in spatial relation with different elements retrieved during the investigation, it is possible to gain a multidimensional representation of the artefacts or sites which include different perspectives and that provide the ground for formulating research questions that go behind the traditional way of looking at archaeological data (see Figures 3.1, 4.5, and 4.7).

Figure 2.7 3D models derived from the volumetric acquisition of a Bronze Age urn burial from Domsten, Sweden. The urn was acquired using a medical CT scan, the meshes are rendered in Cinema 4D. Image courtesy of Claes Ericson (Interspectral), Helene Wilhelmson (Sydsvensk arkeologi). CT scan Acquisition: Marcus Söderberg, (Lund University); 3D data processing and visualisation: Claes Ericson (Interspectral).

2.5 Shaping future archaeological practice

We are led to think that everything that has a presence in the real world can be represented through a 3D model. However, the physical world consists of elements that extend far beyond shape and appearance.

An example of this common to archaeological field practice is the feature's interface (such as the cut of a ditch or pit), which is a crucial element for the definition of the stratigraphic sequence, but which is immaterial, and for this reason, is very difficult to represent using a 3D model alone (Valente et al. 2017:708).

There is no one type of 3D model capable of incorporating all aspects of material culture (no matter the resolution or type, they are always a simplification of the reality), and for this reason, all the forms of 3D visualisation (points cloud, surface, boundary, and volumetric models) should be used synergistically for the representation of 3D space. Boundary models are very efficient for simulating the topological relation between the different deposits encountered during the field investigation or for presenting different interpretations of monuments and sites, while 3D surface models can be more effective for representing context interfaces and for reviewing site morphology. Volume models can be employed to identify distinct elements contained within materials and deposits (structures, artefacts, samples) and to analyse in detail the formation process. The synergetic use of all these different types of models in support of archaeological practice allows representing multiple aspects of the information and provides researchers with the possibility to create a broader and more accurate representation of the archaeological record. By using the concept of "boundary objects" introduced by Star and Griesemer (1989), we can think of 3D models as both concrete and abstract objects, capable of incorporating information from different disciplines and meeting the demands of multiple specialists (Star and Griesemer 1989: 408, 412).

2.6 Summary and conclusions

This chapter presented an overview of the different types of 3D models used to simulate various representations of the archaeological record. It provided a discussion concerning limits and potentials of different types of 3D representations within the framework of archaeological practice, concluding that for a complete geometrical description of archaeological investigation environments, it is necessary to include as many types of 3D models as possible, since each of them represent an approximation of reality. In the next chapter, an overview of the different 3D visualisation platforms currently used in archaeology in support of 3D visualisation and spatial analysis will be introduced and discussed. The chapter will review the capacity of 3D Geographical Information Systems (GIS) to provide archaeologists and cultural heritage practitioners with a digital environment characterised by sustained relations (Janz 2018) that can be used for supporting the process of archaeological interpretation.

3 3D GIS in archaeology

3.1 Introduction

As discussed in Chapter 1, it is apparent that Geographical Information Systems (GIS) have become a de facto standard for visualising and managing archaeological data and linking different archaeological records. When it came to the creation of 3D GIS – although the idea of combining 3D models and GIS was already theorised and widespread (Jones 1989; Raper 1989; Pigot 1991, 1992; Harris and Lock 1996; Gillings and Goodrick 1996; Lock 2001:161, 2003:182; Merlo 2016; Conolly and Lake 2006) – technical and logistical limitations hindered the implementation of well-established pipelines. Indeed, although the term "3D GIS" has been in use since the early 1990s (Goodchild 1991; Reynoso and Castro 1994; Harris and Lock 1996), with some authors trying to move a step forward towards the definition of a "4D"/temporal GIS or TGIS (Castleford 1991), today we are still far from reaching the full potential of 3D GIS. And although there is a wealth of studies engaging a range of aspects of the technology (voxels, DEMs, boundary, surface, etc.), many further improvements are still possible.

Before the spread of 3D surface, boundary and volume (voxel-based) models – as a result of the implementation and spread of laser scanning, geophysics, and image-based 3D Modelling – 3D GIS mainly relied on the use of 2.5D representations. Despite their limits, these models were pivotal for exploring the potentials connected with using the third dimension in archaeology. The drawback of such 2.5D representations lies in their inability to display more than one elevation value per x and y coordinate (de Cambray 1993; ESRI 2018). These limitations are less evident when visualising large portions of the landscape; however, when employed to describe more (geometrically) complex archaeological features, their use tends to be more problematic (see Figure 3.4b). For these reasons, 2.5D representations are not considered "true 3D" because they are incapable of storing discontinuous surfaces, providing an insufficient geometrical description of the objects being represented (ESRI 2018). Some authors have remarked how 3D GIS platforms should provide researchers with the possibility of reviewing the contents and the topological relations within the 3D entities, portraying the 3D GIS as a space capable of managing volumetric models that include features, layers and objects (Van Leusen and Van Gessel 2016:34). Although such a granular level of visualisation has not yet been achieved, a lot of progress has been made in recent years that has allowed 3D GIS to engage with the archaeological space in new ways (Bezzi et al. 2006; Lieberwirth 2008; Opitz and Nowlin 2012; Orengo 2013; Landeschi et al. 2015; Berggren et al. 2015; Dell'Unto et al. 2015; Roosevelt et al. 2015; Merlo 2016; Katsianis et al. 2021; Nobles and Roosevelt 2021), in which multiple new forms of 3D representations were incorporated for solving specific tasks (Lingle et al. 2015; Gavryushkina 2018; Landeschi et al. 2019; Katsianis et al. 2021).

DOI: 10.4324/9781003034131-4

Today, 3D GIS platforms have the potential to map more than just the physical features of the landscape. Intangible aspects of the past, such as sensorial experiences, can also be simulated, promoting the integration of different types of evidence into a 3D model in exciting and unconventional new ways (Landeschi et al. 2015; Polig 2017; Polig et al. 2020).

Wheatley and Gillings described the GIS as the interaction of five main systems: (1) the data entry, which allows importing the spatial data into the GIS; (2) the spatial database, which allows storing topological, spatial and attribute information; (3) manipulation and analysis, which allow data to be transformed; (4) reporting and visualisation, which is responsible for displaying the results of the analysis in form of graphics; and (5) the interface, used for sending instructions to the other systems and optimising the users' workflows (Gillings and Wheatley 2002:10–11). Klinkenberg suggests that in order to be considered fully working in 3D, a GIS should manage all these systems in 3D (Klinkenberg 2016:40). This implies not just the development and implementation of new tools but a shift of practice, where the information is acquired, processed, linked and analysed multidimensionally.

It is only in recent years that archaeologists have started to take advantage of fully working 3D GIS platforms to handle and analyse their own field data (Bezzi 2006; Lieberwirth 2008; Opitz and Nowlin 2012; Orengo 2013; Berggren et al. 2015; Dell'Unto et al. 2015, 2017; Roosevelt et al. 2015; Landeschi et al. 2016; Merlo 2016; Gavryushkina 2018; Katsianis et al. 2021; Scott et al. 2021).

Initially, the lack of development of this promising new approach was attributed to the costs connected with 3D recording and with the unavailability of tools and software designed for 3D visualisation and analysis (Gillings and Wheatley 2002; Merlo 2016:2; Van Leusen and Van Gessel 2016:35). The recent accessibility of advanced 3D acquisition and visualisation systems triggered an intense phase of experimentation, which brought to light new (and more conceptual) obstacles concerning the integration of such tools within the archaeological practice. The extensive use of 3D spatial technology in the field implied a profound revision of several aspects of the field investigation. The introduction of these digital methods within the everyday investigation routines was tedious and initially insufficient for promoting novel explorative and cognitive-investigative approaches (Taylor and Dell'Unto, 2021). This is because the technology used to record and visualise archaeological data affects our understanding of the materials encountered during the investigation (Huggett 2017), and their systematic use in the field can be perceived as replacing more flexible recording techniques with stricter, formalised methods (Morgan and Wright 2018). For this reason, introducing such a new set of tools within our daily practice was far more complicated than initially foreseen.

As previously argued, 3D models are today considered an essential asset in archaeological practice and are used at different scales for representing almost any kind of archaeological information (from landscape to artefacts and from material to immaterial). The rapid spread of 3D acquisition instruments triggered an interesting experimentation phase, which led to the definition of routines for supporting the integration of these methods across the entire investigation process (Callieri et al. 2011; Forte et al. 2012; Opitz and Nowlin 2012; Roosevelt et al. 2015; Dell'Unto et al. 2017; Taylor et al. 2018; Landeschi et al. 2020; Scott et al. 2021). If initially 3D models were mainly used for creating bidimensional documentation to be used within more traditional GIS platforms, nowadays these systems are increasingly employed in the field to gain a better understanding of the progression of the investigation, and to review complex spatial relations at different scales of resolution (Figure 3.1).

Figure 3.1 (Kämpinge) archaeological sequence described by 3D surface models. Image courtesy of the Kämpinge project.

3D GIS is not the only platform used for managing 3D spatial data. Among the different options, researchers have been experimenting (or combining) different software for visualising, querying, and analysing 3D models. Building Information Modelling (BIM) and Game-based visualisation systems (GBVS) were employed in different archaeological projects to address specific research questions. 3D GIS platforms are part of this development, and the recent implementation of effective 3D visualisation solutions, the integration of tools for managing multiple types of 3D models and the construction of new instruments for 3D analysis, underlines the central role of the third dimension within a new generation of GIS science. The main limitation of this platform is the restrictions of the current interface in editing and modelling the data once imported in the geodatabase. The way 3D GIS platforms are evolving will impact future practice, and for this reason, their construction shall come as a result of intense and critical use of this technology in the field. One of the key factors which make 3D GIS a powerful tool for the management of archaeological data is its capacity of including – in the same georeferenced space – a great variety of spatial data, providing researchers with an unprecedented overview of the complexity which characterises any archaeological investigation.

3.1.1 Building information modelling

BIM is a conceptual frame initially created for supporting architectural and engineering studies. It was originally designed to facilitate a more integrated approach between different users and for increasing the quality and production of new buildings (Sacks et al. 2018).

Despite those systems being primarily intended to support modern architecture development, BIM has often been used to analyse historical buildings, and in this case, it is known as H-BIM (Heritage Building Information Modelling). This approach relies on the definition of a semantic structure designed for meeting the needs of a multidisciplinary group of specialists and on the use of 3D volumetric and surface models for describing stratigraphic sequences, historical information, analysis of degradation, or deformations and characteristics of the different elements encountered during the analysis (Continenza et al. 2016). The 3D models created through a BIM can be linked to information such as dating, construction stage, state of conservation, and restoration work providing researchers with a dynamic virtual environment capable of addressing complex questions (Continenza et al. 2016:319).

The introduction of such methods brought a substantial change in the way specialists define and combine new strategies for supporting research in building archaeology. BIM has been used within archaeological studies primarily for reviewing stratigraphic relations, and as a tool for inferring the original shape of standing structures which are no longer visible.

An interesting example of this approach can be identified within the activities of the project "KAINUA. Reconstructing, perceiving, disseminating the lost reality. Transmedial technologies for the Etruscan city of Marzabotto". Here a BIM was employed to simulate the building process of the Tuscan temple of Uni, which was identified during the investigation conducted at the archaeological site of Marzabotto by the University of Bologna (Garagnani et al. 2016).

Archaeological evidence and historical information were used in the BIM for inferring the original aspect of the structure and simulating the temple's construction process (Garagnani et al. 2016:83). Such an approach was already theorised in archaeology at the end of the 90s by Colin Renfrew. In his foreword to the book *Virtual Archaeology* edited by Forte and Silotti (1997), he describes how 3D models and virtual reality would force future scholars to use more structured and logical methods for interpreting the past (Renfrew 1997).

BIM systems are mainly oriented towards modern architecture. These platforms are designed to deal with geometries characterised by standard primitives and equality of proportions. This aspect was considered an obstacle for projects oriented to historical contexts (Continenza et al. 2016:322), but despite these limitations, in the last decade, BIM has had a significant impact on the way historical and archaeological buildings are perceived and studied. BIM has been adopted for supporting post-excavation activities, but it does not include all the types of spatial data usually encountered during a standard archaeological investigation. BIM should therefore not be directly compared with 3D GIS, as they are very different technologies, but the integration of some of the tools offered by this system in the GIS (and vice-versa) is both likely and desirable.

3.1.2 Game-based visualisation platforms

Game-based visualisation platforms have been adopted in several archaeological projects for visualising different types of 3D spatial data. Differently from 3D GIS or BIM, these permit a more flexible design of the graphical user interfaces (GUIs), providing tailored solutions for data interaction and visualisation. Game-based visualisation platforms are naturally designed to build any kind of 3D environment and simulate different phenomena, including virtual characters. These systems have been employed to assess archaeological hypotheses (Frischer and Fillwalk, 2013.), link and explore large repositories of archaeological information (Buckland et al. 2018), and analyse the relations between landscape and architecture (Agugiaro et al. 2011).

The Giza project represents an early example of the use of game environments for linking multiple types of archaeological data in a 3D virtual world. The project was developed by Harvard University and focused on implementing a visualisation system for linking and displaying data sets collected across two centuries of archaeological investigation at the Giza Plateau (Der Manuelian 2013).

The platform includes virtual characters, illustrations, photographic material, and 3D interpretations of objects and structures, which are displayed according to their original spatial position and chronological order (Der Manuelian 2013). This project demonstrates the capacity of such platforms to be adapted and used for testing new forms of interactions with archaeological data, thus catering (within the same system) to a broader range of stakeholders.

Similar aspects were discussed and explored within the Gabii Project (https://gabii.cast.uark.edu/gabiigoesdigital/), an archaeological investigation led by the University of Michigan focusing on the Latin town of Gabii. Since 2009 the project has employed image-based 3D Modelling techniques and GIS for recording and assembling stratigraphic sequences (Opitz and Nowlin 2012). Part of these data were made available through a game-based visualisation platform designed to integrate archaeological datasets, site reports, realistic models and reconstructions (Opitz and Johnson 2016). Both these projects represent successful attempts at using game-based visualisation technology for explicitly addressing the community of archaeologists. Despite their potential, GBVS are mainly used during post-investigation activities and these platforms are often employed to build visualisation systems that address the general public. The possibilities of combining descriptive data within fully reconstructed worlds could help future archaeologists gain more resolute interpretations of the past (Figure 3.2).

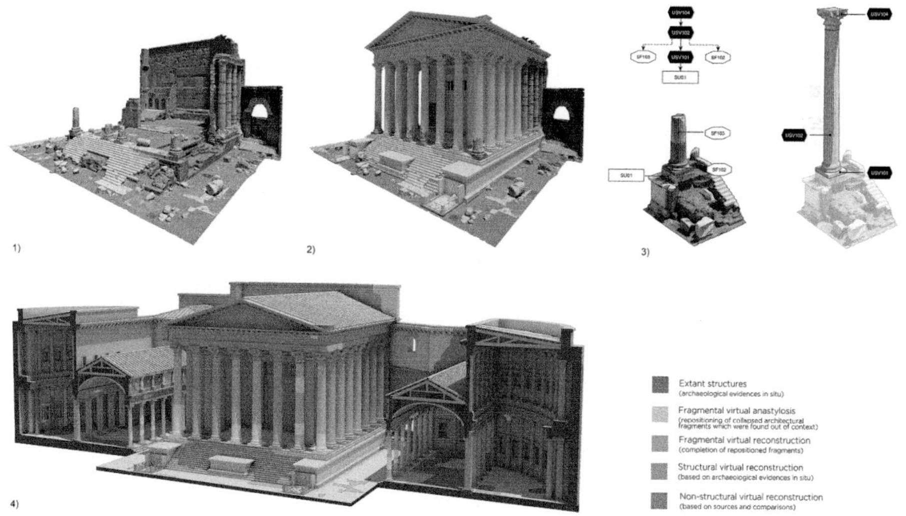

Figure 3.2 (a) 3D model of the Forum of Augustus, developed by the National Research Council of Italy CNR ISPC in collaboration with the Museo dei fori Imperiali and VRTron as part of the EU project REVEAL. The virtual interpretation of the structures is made using the reality-based models as geometrical references. (b) Example of 3D interpretation using the extended matrix approach (Ferdani et al. 2020). Image courtesy of Daniele Ferdani.

3.1.3 Integrated 3D GIS-based solutions

As stated in paragraph 3.1, three-dimensional GIS platforms have existed at least from the 1990s/early 2000s, with commercial and open-source solutions available and used by a very active community of practitioners. Nevertheless, to fully understand the problematic nature of most of these products it is important to outline the broad and often blurry definition of what a 3D content or a 3D visualisation actually is.

In computer graphics, 3D visualisation is a method that employs 3D contents (or 2/2.5D or textual contents) displayed in a Cartesian 3D space (defined by x, y, z coordinates) for extracting meaningful information, communicating a message, clarifying spatial or topological relations, or for better illustrating phenomena or processes, through the use of interactive graphics and imaging (Kaufman et al. 1993). On a landscape scale, GIS practitioners have been using 2.5D data visualisation tools to represent DEM-related datasets, featuring the topography of the study area, and providing a basis for further spatial analyses (Lake et al. 1998; Gillings and Wheatley 2002). However, due to the limitations of visualising more geometrically complex datasets, several solutions have been explored for the visualisation of 3D data.

Surface, boundary and volume representations have often been managed in the GIS using external or additional visual solutions, thus decreasing the possibilities of contextualising 3D data within the rest of the investigation's documentation.

A significant example of this phenomena can be found in ESRI Arcview; the system relied on an external visualisation tool that could visualise (in the same virtual environment) TIN, shapefiles and data from computer-aided design (CAD), as well as being able to perform line-of-sight analyses for creating 3D visibility maps (Ormsby and Alvi 1999). The system had a large diffusion during the 90s but the fact that the software could only visualise 3D data in an external window (or module) with very basic functions prevented a deeper interaction with the 3D records, significantly reducing the impact of the 3D records in the process of data interpretation. More advanced solutions towards the visualisation of 3D data were explored using the GIS software GRASS; this was one of the first platforms that could manage 3D volume data in an open-source environment. During the last decade this platform has been used to develop several experiments aimed at representing and exploring archaeological deposits using volume pixels. The software employs dedicated plugins (https://grass.osgeo.org/grass78/manuals/addons/r.vol.dem.html) to represent sub-surface information related to spatial phenomena or for representing and exploring the stratigraphic relations among the different deposits encountered during field investigations (Merlo 2004; Bezzi et al. 2006; Lieberwirth 2008).

Once produced, volumetric (voxel-based) data are visualised using external software (such as nviz or Para View) that allow running more sophisticated analyses (Bezzi et al. 2006). As pointed out by Lieberwirth, the combination between GRASS GIS and Para View presents several advantages, such as simulating site taphonomy or erosion, reconstructing the stratigraphic sequence in chronological order, or comparing and analysing the different archaeological deposits encountered during the investigation (Lieberwirth 2008:92–93).

A drawback is that these platforms rely on different visualisation systems (or modules) for the visualisation of the datasets recorded (or produced) during the investigation. In fact, splitting the visualisation across several platforms is not ideal when dealing with multiple sets of spatial data, as it could prevent the identification of patterns only visible when combining all the information in the same view.

More recently, several solutions have been proposed by different commercial and open-source platforms in order to increase the integration of 3D records in the GIS. ESRI ArcGIS entrusted the visualisation of 3D data to ArcScene, a visual solution oriented towards a 3D holistic data representation. Although the 3D datasets managed by ArcScene were initially very limited, the later implementation (and development) of multipatch files meant that it could include and manage 3D surface and boundary textured models in the geodatabase (see Figure 4.2). The open-source platform GVSIG was one of the first GISs that attempted to integrate 3D georeferenced surface textured and boundary models in a 3D visualisation environment, demonstrating the advantages of visualising a larger number of 3D datasets in one 3D view (Bezzi 2012). Although the development of this platform was stopped shortly after the implementation of this interesting function, this approach clearly demonstrated the potential of including a larger set of 3D data in spatial relation with more traditional archaeological graphic documentation. This option was then further implemented by ESRI, which recently started investing considerable resources into creating GIS platforms oriented towards 3D visualisation and analysis (e.g. ARCGIS PRO or CityEngine).

Another interesting integrated solution is the connection between QGIS and Blender software through the use of a dedicated add-on, allowing users to export shapefiles from QGIS and visualise them in Blender's 3D visualisation environment. The main issue is still the difficulty of performing typical GIS-based tasks in a 3D modelling software. At the moment, full 3D GIS solutions are one of the most viable technologies for performing both GIS and 3D modelling analyses in a single space. However, the possibility of employing powerful editing tools for manipulating the spatial information directly in the 3D visualisation space (e.g. generating boundary models, visualising vertical surfaces such as stratigraphic sections, or generating new interpretations in the form of 3D vector layers) proved to be an important element for reviewing the archaeological space. The combination of such systems encourages the formulation of advanced and well-structured forms of 3D interpretation, and it is often adopted within the frame of projects focused on virtual reconstructions or simulations.

A more integrated solution designed for supporting the work of field archaeologists is represented by the later version of Intrasis (https://www.intrasis.com/news/), a commercial GIS software developed by the archaeological company Arkeologerna and based on ESRI ArcGIS architecture and PostgreSQL database functionalities. The current 3D version of the software allows users to integrate 3D vector objects within the 3D canvas visualisation environment and can be used to edit 3D polylines that are connectable to the DB structure of the software. The 3D models imported in Intrasis are well managed by the visualisation platform, but can't be stored in the geodatabase yet, thus decreasing the possibilities of using the software for performing more advanced analysis on the models. This platform represents an excellent example of migration of practice from a 2D to a 3D environment, and its use has provided important indications concerning the limits and potentials of using 3D records in support of archaeological field practice.

The latest release by ESRI (ArcGIS PRO) can manage various types of 3D models in the same virtual scene (e.g. volume, surface, boundary ones), at different scales of resolution and in spatial relation with more traditional GIS datasets. The 3D models employed during the project can be stored in the geodatabase with the rest of the documentation and can be used for extrapolating additional records (see BOXs 4.3–4.5, BOX 7.2). Differently from other solutions (e.g. Intrasis) this platform is designed for serving a broader range of GIS applications and for this reason it requires a high level of customisation before being used in the field, and in its full 3D extension requires significant licensing costs.

3.1.4 3D web GIS solutions

Web 3D GIS platforms are technological solutions employed for storing, managing, displaying, and analysing spatial information through the World Wide Web on a 3D visualisation canvas. These can be used by any user with access to a stable internet connection, without installing any specific Desktop visualisation software. These platforms have a strong impact on archaeological practice because they allow mapping, visualising, and analysing archaeological features using web collaborative environments and can therefore reach a much larger number of specialists, almost in real time. More than any other systems, 3D web GIS platforms open up archaeological records to the world, transforming the investigation from being an isolated phenomenon to a digital, multivocal event.

Archaeological records stored on a Desktop 3D GIS can only be used by users or institutions with access to those platforms, reducing the opportunity for other practitioners (e.g. scholars or students) to employ the datasets stored in the geodatabase for testing new hypotheses or assessing new interpretations.

The lack of interoperability of Desktop 3D GIS (Jensen 2018: 32–33) and the more recent development and spread of 2D-3D web GIS platforms represent an encouraging alternative for exploring and experimenting with new forms of data interaction through the web (Galeazzi and Richards-Rissetto 2018:2).

A few Desktop 3D GIS platforms can publish simplified versions of 3D GIS projects. However, the use of these online visualisations is still more oriented towards divulgation rather than interpretation and analysis. Among the few examples that integrate more advanced visualisation, system architecture and analytical tools, the Maya Arch 3D project and Archeo 3D viewer are customised to include all aspects of archaeological investigation (from data acquisition to data publication). Maya Arch 3D (https://mayaarch3d.org/en/) is designed for hosting and analysing archaeological records acquired across several field investigations. The online system focuses on the eighth-century Mayan kingdom and is one of the first examples in which 3D models are used in an online 3D GIS platform. This provides the user with a virtual environment that can be used for assessing archaeological interpretations and reviewing different types of records. The system combines WebGL, Javascript as well as PostgreSQL (Agugiaro et al. 2011; Auer et al. 2014), and it is considered an essential reference within the community of practitioners.

Archeo 3D is a system designed for the real-time exchange of archaeological records among archaeologists and specialists working on site. The system is flexible, dynamic, and grants an easy implementation and management of 3D models within a visualisation environment customised for supporting archaeological practice. It includes several powerful analytical tools that facilitate the use (or reuse) of records related to the ongoing investigation (Jensen 2017, 2018). Archeo 3D is designed for serving multiple types of archaeological environments and it promotes new forms of archaeological documentation (Jensen 2018:40). By documenting "events" (or specific groups of records) rather than single elements, the system explores new forms of documentation that are designed to promote and increase multivocality and reflexivity.

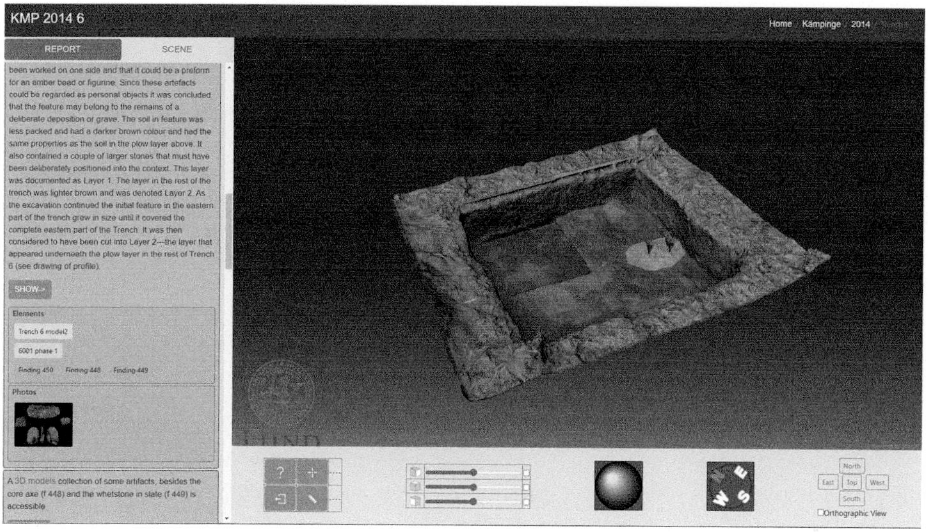

Figure 3.3 Interactive Reporting System (IRS). The system is available online and includes data collected across four years of investigation. IRS is customised to support field investigations in Sweden (Derudas et al. 2021).

Other types of 3D web platforms for the publication of archaeological data have recently been introduced as part of various research projects. All these share common aspects of archaeological practice and data reuse (Opitz and Johnson 2016; Sullivan 2020), and provide students and scholars with dynamic and interactive 3D web solutions that can be used for describing archaeological records and activities undertaken across multiple archaeological investigation campaigns (Derudas et al. 2021; Derudas and Berggren 2021). Among the different platforms recently presented, the Interactive Reporting System (IRS) represents an advanced case of a 3D Web platform designed to promote deep interaction with the archaeological records stored on the web (Derudas et al. 2021) (Figure 3.3).

The collaborative aspects connected with 3D Web GIS platforms are gaining momentum and will certainly be included in any future 3D data management platform.

3.2 Boundary, surface and volume 3D representations

Today it is possible to rely on several kinds of 3D models, each one characterised by different limits and potentials. By using examples from the literature, this subchapter will review which 3D data are mostly used within 3D GIS platforms.

Over the years, different types of 3D data have been employed for representing different types of archaeological finds: Total Station or GPS were often used for 3D recording artefact distribution or for generating 3D excavation plans (Barceló et al. 2003; Losier et al. 2007), orthoimages draped on 2.5D raster DEMs have been used for displaying context interfaces (Katsianis et al. 2008), while archaeological deposits were represented in the digital space by boundary or volume (voxel) models (Merlo 2016).

Using 2.5D orthoimages was an effective solution (probably the most effective, before the spread of laser scanning and image-based 3D Modelling) for representing context interfaces and for reviewing the spatial relations between different materials. However, the low resolution of these models (due to the use of total stations for creating a grid of 3D points for generating a TIN) and the interpolation problems related to the use of 2.5D surfaces decreased the possibilities of reaching a clear 3D visualisation of space, especially when tackling more complex geometries (Figure 3.4).

Voxel representations provide a very detailed description of archaeological deposits; however these require large amounts of storage space and are not as accurate at representing entity shapes (specifically if they have curved surfaces) (de Cambray 1993:343).

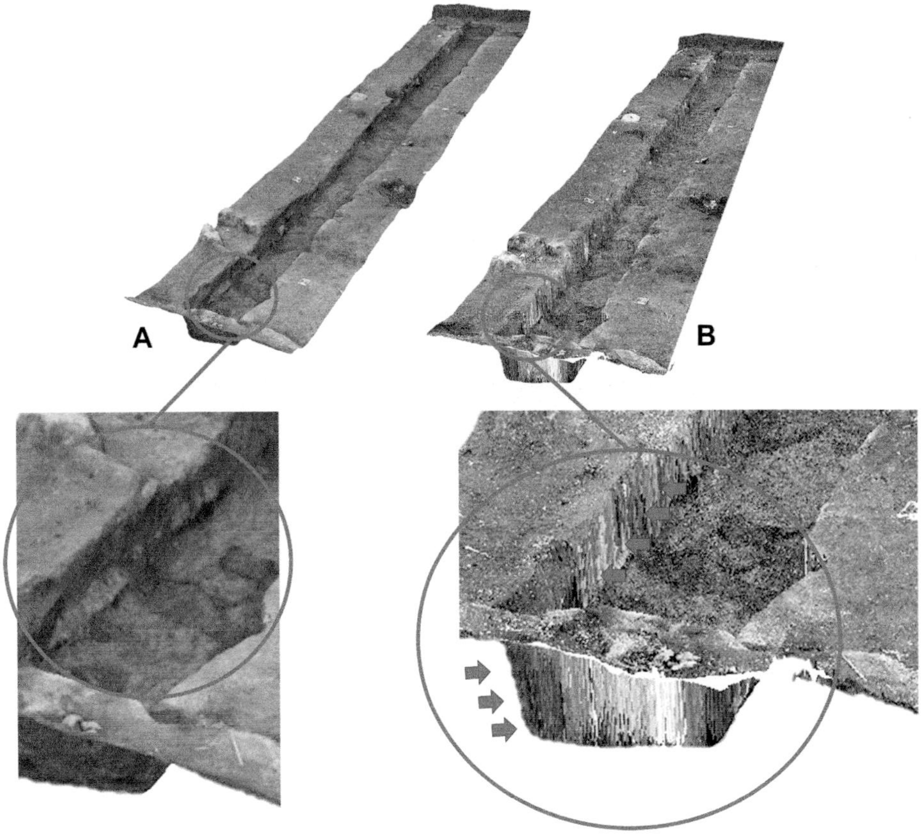

Figure 3.4 On the right side (B) orthoimages draped on the 2.5D model; on the left (A) 3D surface model generated using image-based Modelling. The geometry is derived from the same 3D point cloud processed with Metashape 1.7 Agisoft. Both models are displayed using ESRI ArcGIS. The 3D surface models (A) can manage multiple elevation points for x and y coordinates, providing a proper 3D geometrical description of the model in all its parts. Despite the apparent similarity of the two visualisations, model (B) is a 2.5D representation that cannot display more than one elevation value per x and y coordinate, preventing a correct visualisation of the vertical surfaces of the section. 3D models Nicolò Dell'Unto, photographic acquisition Sjoerd van Riel and Giacomo Landeschi.

More simple boundary representations (e.g. those derived from the combination of multiple 3D point clouds) permit extracting information concerning the volume, and although they often result in intense data processing, they seem to be a very efficient solution for visualising and revising topological relations among different deposits encountered during the investigation (see BOX 4.5, BOX 7.1). 3D surface models have been employed for recording and visualising contexts and features detected during field investigations and for documenting ongoing excavation activities. Despite the impossibility of using this type of models for representing volumetric characteristics of the information, they have nevertheless been useful for field investigations due to several factors: the hyper-realistic aspect of the representations (especially when using image-based Modelling 3D techniques for their production), their versatility (they can be further combined to produce the boundary of voxel-based models), and the possibility of producing and using these models at the trowel's edge (for supporting a real-time interpretation of the archaeological contexts and materials in the field). Moreover, it is also worth noting that this type of data is often used as the primary source for generating all the other 3D representations currently in use for representing archaeological records (e.g. boundary and voxel-based models).

3.2.1 Volume representations

At the end of the 1990s and during the first decade of the 2000s, several experiments involving voxel models were carried out for supporting visual analysis and archaeological field interpretation (Harris and Lock 1996; Bezzi et al. 2006; Lieberwirth 2008; Merlo 2016). Voxels are volumetric elements particularly suited for representing environmental or geological data and are usually generated during post-excavation activities by filling the empty space between two boundaries (usually two context interfaces) with identical non-intersecting geometrical entities or by interpolating point clouds (de Cambray 1993:342; Bezzi et al. 2006; Lieberwirth 2008).

Voxels are often employed for clarifying the relations between the different strata encountered during the investigation and for performing more accurate geostatistical analysis (Lieberwirth 2008:87). Despite the encouraging results, voxel models were never considered (or adopted as) a standard in archaeology. The use of this type of 3D representation, however, has begun to impact archaeological investigations that use instruments that can record and assign different values to the different deposits perceived underground (see Re et al. 2015). One of the promising areas of application of voxel-based models is geophysics. Indeed, among non-destructive methods of field investigation, geophysical prospection represents a core technology that has been widely used in landscape archaeology for investigating buried structures and sub-surface deposits, and there is a vast literature describing its application in various contexts (Clark 1968; Campana and Piro 2008; Gaffney 2009; Witten 2017). Information derived from geophysical prospection has been employed to improve a site's interpretation and to detect new features not visible on a surface level. For instance, the integration between geophysical and airborne remote-sensed data has proven to be a valid support for investigating archaeological landscapes (Larsson et al. 2015; Verdonck et al. 2020). As a further step, datasets derived from this type of prospection have served as a geometrical reference to make a 3D reconstruction of landscape elements (Klein et al. 2012). Nevertheless, so far geophysical data and in particular those acquired through Ground Penetrating Radar (GPR) have rarely been really exploited to their full potential, namely conveying 3D information about sub-surface archaeological deposits. In a recent article, Verdonck et al. (2020) point out that the use of GPR data in GIS is "limited to the display and overlay of horizontal 2D time-slices" [representing different layers of the deposits] (Verdonck et al. 2020:716). What is missing, according to the authors, is the possibility of visualising in a GIS environment the result of GPR acquisition, either in the form of a 3D point cloud (where each point corresponds to a single sampling unit as a result of the sensor signal acquisition) or as an interpolated voxel model in which the 3D space around each sample unit is assigned with a specific value, defined according to different possible interpolation algorithms. As will be described in Chapter 7, Desktop 3D GIS solutions available today provide users with some interesting functions for handling 3D point clouds and for transforming data sets of points into volumetric units capable of describing a minimal space assigned with a value that corresponds to the original sampling unit (see BOX 7.2). When it comes to the visualisation and the description of large quantities of point datasets, the Empirical Bayesian Kriging 3D function (ESRI 2020) can, for example, interpolate the point cloud into a volumetric model defined by voxels. Although voxel data models are not a novelty per se, (see paragraph 3.1) the possibility of integrating different types of 3D data models into the same GIS environment (including volume, surface, and boundary ones) represents a remarkable turning point that will have a significant impact on the discipline. This will include the possibility of putting geometrically complex textured surface and boundary models of the site/landscape in the right spatial relation with sub-surface models, defined by the volume of the "anomalies" detected by GPR acquisition. This will also provide a quantitative assessment of the amount of archaeological information lying underground.

3.2.2 Volumetric analysis of non-material phenomena

Another interesting field of applications for a volumetric 3D GIS relates to the analysis of non-material phenomena, such as visibility or light, which characterise the reality of the past and which have been often recalled by scholars in an attempt to define new ways of engaging with the study of ancient space. Recent scholarship (Gillings 2017) shows that archaeological GISs can be effectively employed to investigate concealment and seclusion as important aspects characterising human choices in a certain landscape or architectural space. Although visibility studies have long played a major role in GIS-based analysis, scarce attention has been devoted so far to what we could term "invisibility". By taking this perspective, it would be possible for archaeologists to investigate the motivation behind otherwise unexplained phenomena, such as the case of the "minilιths" of Exmoor, UK, prehistoric monuments built with small stones whose hidden location in the landscape suggests they were deliberately intended not to be seen (Gillings 2015a, 2015b). By applying computational methods based on raster-based viewshed or vector-based line-of-sight (LOS) analysis it is indeed possible to measure the level of visual exposure associated with a predetermined set of target features in order to determine how visible these objects/monuments are from a series of hypothetical observing points. Still, the main limitation of most of the current GIS-based approaches to visibility lies in the fact that the analysis is performed on 2.5D data with only a few attempts in existence that explore the potential of a fully 3D approach (Landeschi et al. 2016; Polig et al. 2020). As will be described more thoroughly in Chapter 6, one of the main advantages of opting for a volumetric data model is the possibility of visualising and quantifying visual space in its continuity and measuring the level of visibility/invisibility associated with the space between features. This so-called "emptyscape" is often neglected from current archaeological discourse, but should instead be analysed to provide a deeper insight into the complexity of the spatial continuum (Campana 2016) and thus allow us to overcome the traditional, dualistic site/off-site approach to landscape archaeology. As an additional point, volumetric GIS-based analysis would be beneficial for investigating phenomena such as natural and artificial illumination, whose agency in relation to ancient architectural space, as well as prehistoric dwellings, has been highlighted by scholars in recent years (Earl et al. 2010, 2016; Papadopoulos et al. 2015).

3.2.3 Surface and boundary models

With the development of more powerful computational machines and instruments for 3D recording (Laser scanning and image-based 3D Modelling), archaeologists began to experiment with the use of 3D resolute textured surface and boundary models for documenting contexts, features and artefacts encountered during the field investigation, and for recording and visualising (in 3D space) geometrical characteristics and spatial relations among archaeological structures. The novelty was not the type of 3D models (3D surface and boundary models were already produced and used in archaeology since the end of the 90s) but rather the new set of affordances offered by the quality and resolution of this new generation of datasets. These specific 3D data significantly improved the accuracy and level of detail of the information recorded in the field, opening interesting new scenarios in terms of representation and simulation. An important step for integrating these datasets within the investigation process was the establishment of a link between the 3D surface and boundary models and the more traditional datasets usually produced during the investigation.

For this reason, several experiments were conducted exploring the possibility of including and managing these 3D datasets within a spatial geodatabase. Most of the archaeological projects engaging with 3D textured surface and boundary models were developed using ESRI ArcGIS (www.esri.com). This platform provides multiple solutions for visualising and analysing 3D data and can count on a large community of developers and practitioners. One of the first attempts to integrate in ArcGIS 3D surface models derived from 3D recording techniques was carried out during the study of the abbey of Niedermunster, France (Koehl and Lott 2008). This project provided an early critical assessment of the limits and potentials of the data formats (e.g. mutipatch files) developed by ESRI for visualising surface and boundary models.

Multipatch files were designed for the boundary representation of 3D objects (ESRI 2012), and these have been especially important for experimenting with advanced 3D visualisation of archaeological records (Opitz and Nowlin 2012; Dell'Unto et al. 2016). In the last decade, multipatch files have been tested as part of various research projects for managing geometrically complex 3D models in a geodatabase (ESRI 2008). Despite their initial limitations in managing high-resolution 3D models (Verhoeven 2017:1016, 1017), they were employed by specialists for addressing specific archaeological research questions (Polig 2017; Lingle et al. 2015; Wilhelmson and Dell'Unto 2015; Landeschi et al. 2016).

These attempts have allowed users to partially overcome some of the limitations connected to the traditional use of GIS, promoting critical discussion and fostering new methodological and theoretical perspectives. Today, different software solutions allow importing high-resolution 3D Models in OBJ format, solving previous issues related to the visualisation of the models in a georeferenced space (see BOX 4.3) (Figure 3.5). These recent improvements of the software address most of the critical points raised in the past concerning the use of 3D GIS as the main tool for 3D data management and visualisation in archaeology.

The possibility of visualising 3D representations of archaeological units encountered during the field investigation in a 3D GIS platform provides archaeologists with the possibility of using the 3D space for experimenting with different forms of knowledge production (see Chapter 4). Complex archaeological features or standing structures, often removed during the investigation process, can now be used to populate the 3D space and review and analyse different kinds of evidence encountered across different field campaigns in spatial relation to them (Figure 3.1). As previously mentioned, 3D surface models are increasingly adopted in GIS as the primary form of documentation

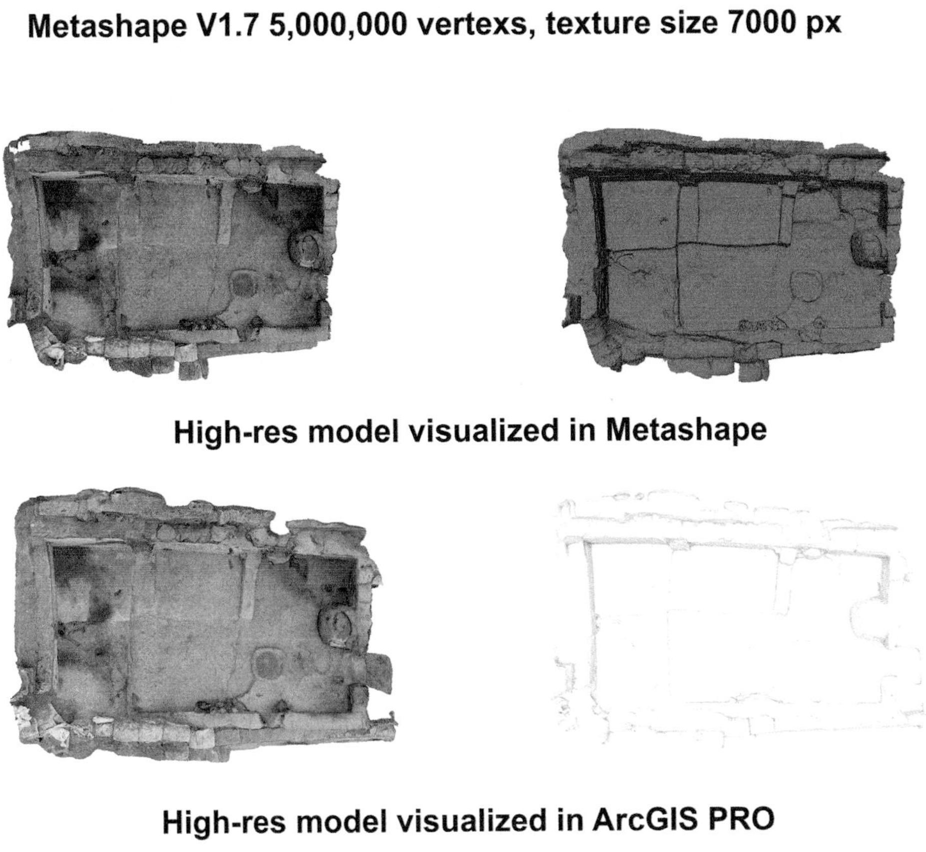

Figure 3.5 3D model of building 80, at the archaeological site of Çatalhöyük, Turkey. The model displayed in the figure was generated using Agisoft Metashape 1.7. The model's geometry is composed of 5,000,000 vertices with a texture size of 7,000 pix. Once created, the model was scaled, georeferenced, exported as an.obj file and imported directly into ArcGIS PRO in full resolution. Courtesy of the Çatalhöyük Research Project.

and are used (or combined) to produce different types of graphic representation such as orthoimages, boundary and voxel models (Katsianis et al. 2021).

Various authors started discussing this approach already in traditional GIS environments (Wheatley 1995; Llobera 1996; Gillings and Wheatley 2000; Lock et al. 2014), and the new opportunities created by the development of 3D visualisation within GIS underline the importance of such an approach for reaching a more comprehensive understanding of past events.

3.3 Implementing 3D geometrically complex models in GIS: an *insula* from Pompeii – a case study

This section presents the results of experiments and tests conducted in order to import geometrically complex 3D models into a GIS. The case study relates to a Pompeian building and will be used to illustrate the work pipeline for importing 3D boundary models derived from different acquisition techniques in a GDBMS. The case study shows the potential of handling 3D data information in a structured geodatabase management system, with the possibility of relating it to multiple data sources (digitally scanned plans and section drawings, photographs, etc.) in a coherent way.

A new generation of 3D GIS platforms provides users with multiple solutions for the visualisation of geometrically complex 3D models. However, as previously mentioned, moving from a 2D to a 3D spatial environment implies a shift of practice in which all five systems which compose a GIS (data entry, spatial database, manipulation and analysis, reporting and visualisation and interface; see paragraph 3.1) are revised and structured for including a multidimensional 3D representation of the archaeological space.

The most challenging part of this transition was probably the development and adaptation of more traditional practice for these new visualisation tools. The results of this transformation started being visible only after developing several field projects where different types of models were produced, implemented, and analysed for addressing

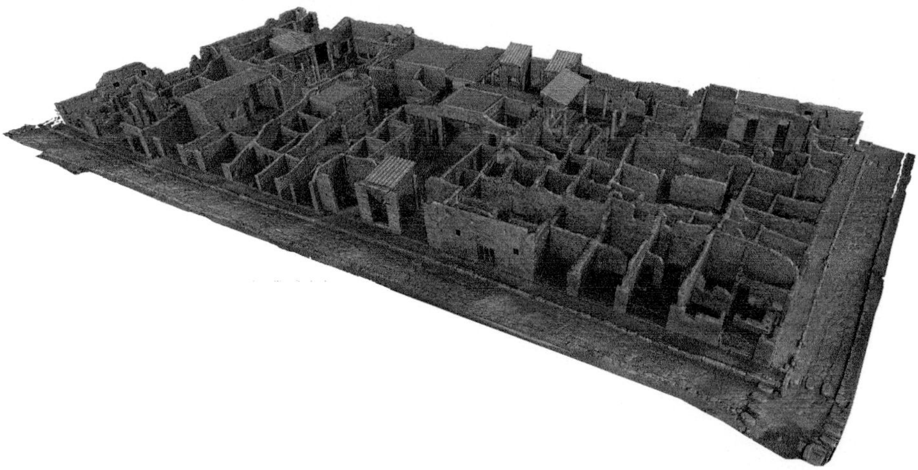

Figure 3.6 3D model of the insula V1, Pompeii. The Model was acquired by combining Laser scanning and image-based 3D Modelling techniques. Once created, the model was georeferenced and imported into the 3D GIS platform used during the investigation. Image courtesy of the Swedish Pompeii Project.

specific questions. This process (which is still going on) was fundamental for developing a new approach to data collection, data management, and data analysis. One of the first and most challenging projects in which we were "forced" to strongly revise our GIS practice was the Swedish Pompeii Project (SPP), where an entire Pompeian house has been digitally acquired and transformed into a 3D model (Figure 3.6), split into different analytic units (rooms), and imported into a bespoke ESRI ArcGIS geodatabase management system (GDBMS).

Results of this experiment were published in Dell'Unto et al. (2015) and represent one of the first examples of systematically importing and managing a significant amount of 3D information in a coherent, relational GIS structure which has also been connected to the project website.

For the reader's clarity, the description of the process starts with the definition of the relational database scheme at the core of the system and continues with a step-by-step description of the georeferencing of the 3D boundary models. Despite there now being multiple ways of geolocating 3D boundary/surface models in a GIS, the following example is still relevant, especially in those cases where the original 3D models lack ground control points and need to be spatially adjusted directly into a GIS based on references provided by additional layers of information (raster/vector basemap). The following case study also explores the problem of contextualising and reusing archaeological records available from older projects. As a concluding part, we describe the connection between the models and the project website's related information, pointing out some basic functionalities that the 3D GIS environment can enable. Whether the examples are based on a specific case study, the workflow here defined can be extended and applied to all archaeological studies based on 3D models derived from laser-scanning acquisition.

3.3.1 Designing a geodatabase structure

In 2000, the Swedish Institute in Rome started a research project aimed at studying and clarifying the relationships among the archaeological structures of the insula V 1 in Pompeii. Due to the complex stratigraphic and spatial relations of the buildings that characterised the insula, several documentation techniques were employed. The data collected in the field were organised and published on the project website (www.pompejiprojektet.se), through which it is possible to access the complete documentation produced by the project team so far. The data are accessible through detailed text descriptions, photographs, orthoimages, maps, sections, and reports. (Figure 3.7).

As a result of a research collaboration between Lund University and the National Research Council of Italy CNR-ISTI, during the fall of 2011 a new project was established, focusing on acquiring a high-resolution model of the entire insula V1 (Dell'Unto et al. 2013b). The use of these technologies permitted generating a spatially accurate documentation of the structures, for reaching a broader and deeper view of the relations between the different elements that characterise the buildings (Dell'Unto et al. 2013a).

The 3D models were implemented in a 3D GIS platform and visualised in spatial relation with part of the documentation available on the website. The system was designed for supporting the archaeological interpretation of the Insula V1 and for providing archaeologists with new tools for supporting interpretation and analysis (Dell'Unto et al. 2015).

As a part of the SPP collaboration, a 3D GDBMS was required for managing the 3D models acquired through laser scanning during the previous field campaign and relating them to the existing site documentation produced by project team members working in the Pompeian city block.

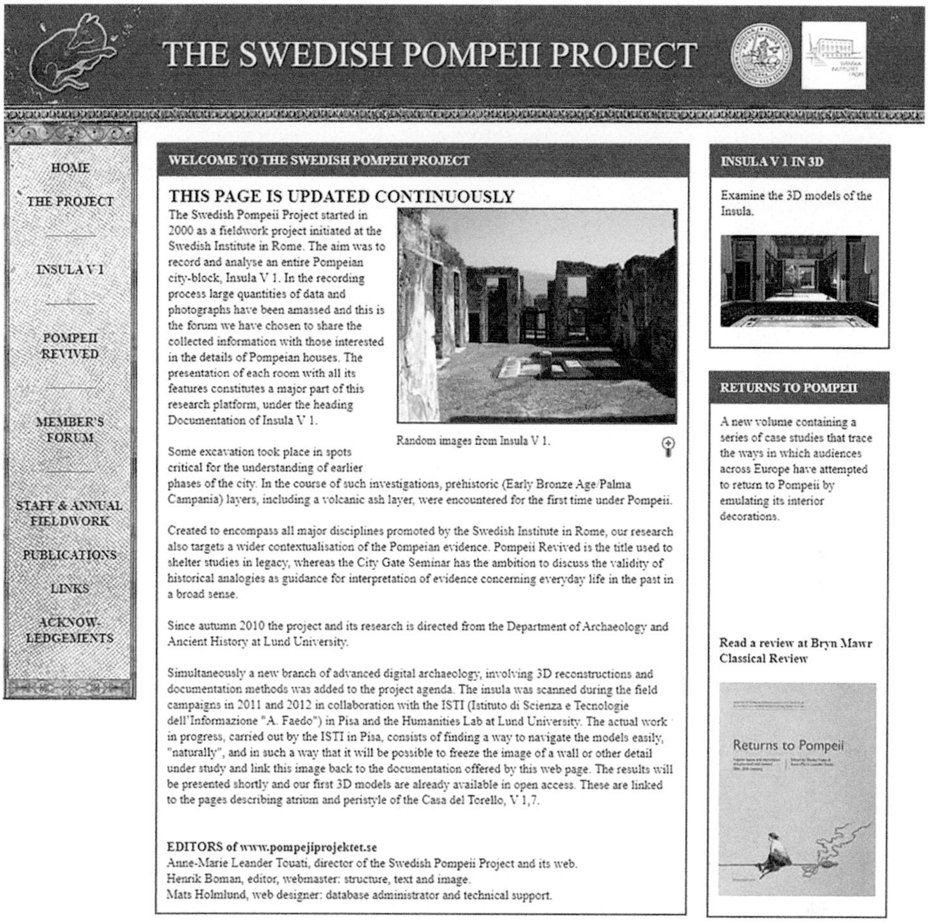

Figure 3.7 The Swedish Pompeii project website (www.pompejiprojektet.se).

The geodatabase structure has rapidly become the native data format for managing spatial and non-spatial information in ESRI-based platforms, with 3D surface/boundary models as pivotal elements included among the different types of sources that can be imported and stored.

To date, the geodatabase standard represents one of the most valuable solutions for handling geometrically complex and textured 3D boundary models in a GIS environment. Interestingly, the GDBMS structure allows users to set up a relational structure of geographical and non-geographical datasets in which data storage functionalities are implemented, along with tools designed to maintain data integrity, to define topologies, and to create relationships between feature classes. In other words, GDBMSs allow practitioners to integrate spatial and non-spatial information and perform traditional GIS tasks within a framework with pre-defined rules relating to the data they contain. Among the available solutions, the "file geodatabase" format resulted to be the most suitable management system for the specific needs of the SPP. It consists of an almost limitless storage space in terms of overall size and a maximum of 1 TB for every single table or feature class (https://desktop.arcgis.com/en/arcmap/latest/manage-data/

administer-file-gdbs/file-geodatabase-size-and-name-limits.htm). As previously mentioned, a multipatch is basically a set of patches storing different types of information including texture, colour, transparency, and geometry. Multipatch feature classes work as well as shapefiles and are particularly suited for users who want to handle models involving a significant amount of triangles and want them to be visualised with a texture on top. In addition, enclosed 3D boundary models allow users to perform topological operations that are typical of vector graphics, such as contains/contained, or buffering (see Chapters 4 and 7). This aspect makes them particularly suitable not only for visualisation tasks but also for data analytics. In addition, the possibility of generating and handling full-3D shapefiles allows users to perform advanced editing operations in spatial connection to a multipatch feature class such as (1) drawing the deposits removed during the excavation in 3D, (2) outlining the archaeological contexts visible on the 3D models' surface (see BOXs 4.3 and 4.4), or (3) managing 3D boundary representations of the archaeological deposits derived from the interpolation of multiple sets of 3D point clouds (see BOX 4.5).

As a first step in the implementation of the dataset, only a portion of the entire city block (insula) acquired during the field campaign was imported in GIS, namely the southernmost part of the double atrium house belonging to Caecilius Iucundus (Dell'Unto et al. 2015). The relational structure of the geodatabase consisted of four main entity categories: (1) objects, (2) tables, (3) field descriptors, and (4) hyperlinks. Objects meant all entities with any spatial connotation, and they were, in turn, divided into three main categories of data:

- a) 3D surface models of the acquired building.
- (b) Raster maps derived from a digital scan of old excavation drawings (sections and plans) including topographical maps of the study area used as references.
- c) Any new informative layers produced as a result of on-site digitisation.

3D boundary surface models (a) were obtained as a result of data post-processing applied on a raw point cloud created from integrated scans performed during fieldwork (Dell'Unto et al. 2015). To make the dataset match the descriptive structure of the documented buildings featured on the SPP website, the whole model of the house of Caecilius Iucundus was split into single room models, each one representing one of the rooms belonging to the house and marked by an alphabet letter.

Concerning the basemap (b), this was made up of reference maps employed to adjust the location of the 3D models according to their geographical position, and ranged in scale from 1:1000 to 1:20. They included a general plan of the archaeological area of Pompeii, plus more accurate plan drawings created by team members of the SPP. In order to improve the local accuracy of the 3D models in relation to their original height, a Digital Terrain Model was derived by interpolating the value of the elevation points measured during the different field campaigns with a total station survey and reported on the digitally scanned map used a spatial reference for the models. On a general level, this local DTM was adjusted in order to be spatially consistent with the information described by the Shuttle Radar Topographic Mission (SRTM) global DEM featuring a spatial resolution of 30 m.

As for the layer "c", this was intended for any information produced during future fieldwork and was created in order to fulfil specific documentation/analysis needs, including structural degradation, chronological analysis, or feature mapping.

Concerning non-spatial data, the first level of information consisted of table datasets, that were linked to room objects through a one-to-many database relationship. Each room had *n* number of "structure" tables related to different architectural features defining that specific room space (such as north wall, south wall, or floor). Each structure entity was, in turn, connected to multiple photograph and drawing tables, containing a description of field documentation material related to that specific structure. All the table entities were characterised by one or more attribute fields that were designed to match the documentation entries described on the previously mentioned project website (Dell'Unto et al., 2015) (Figure 3.7).

3.3.2 Georeferencing 3D surface and boundary models containing no spatial reference

GISs are, by definition, designed to handle geospatial information in which datasets are collected and assigned with geographical coordinates visualised in a Cartesian canvas defined by x, y, z coordinates. Georeferencing is the process that allows users to provide objects from the real world with geographical coordinate information. All the available GIS products, whether commercial or free/open-source, have dedicated tools for geolocating different types of datasets, including raster and vector files. When it comes to 3D models in ArcGIS, the 3D Analyst extension allows users to import and transform different data formats into multipatch feature classes. Since the 3D models of the house of Caecilius Iucundus did not contain any spatial reference, the best option for georeferencing it was to import each single room model into the GDBMS and then apply some roto-translation functions in order to make it match its original position in the geographical space. This was done by using a previously imported map as reference, which featured the plan drawing of the house. As has been described above, a local DTM was then produced to provide an accurate reference surface for adjusting the elevation of the house according to the real-world values.

BOX 3.1 Georeferencing a 3D model using the roto-translation functions

The methodology described in this box is developed using ESRI ArcGIS Desktop/PRO; the 3D models imported in the GIS were acquired using Laser scanning; the point clouds were post-processed using Meshlab for generating the mesh and Blender for projecting the textures. The maps used as spatial reference were produced by the SPP and the DEM was generated using raster data from the SRTM global DEM featuring a spatial resolution of 30 m.

The georeferencing method presented in this BOX allows users to import a correctly scaled surface/boundary model into GIS and place it in its correct spatial location based on the reference provided by the x, y raster map and lifted up to the right z location, based on the Digital Terrain Model (DTM) derived from the total station points measured in the field. The workflow consists of the following steps:

(Continued)

The georeferencing for this case study included several additional steps:

First, all of the height points based on the values derived from the total station survey are drawn on a reference map (**1**) (after Ezequiel Pinto-Guillaume). A DTM is obtained by interpolating the height points previously digitised and used to set the reference map at its actual height (**2**). Each 3D model is imported and, in editing mode, is moved with a delta tool along the x, y, z axes. By using the reference provided by the plan the model is placed in the correct x, y location. Then, the model is "lifted" and placed at the right elevation based on the reference surface provided by the above-mentioned DTM (**3**). The same operation is repeated for multiple rooms that are moved and attached to one another in a topological order enabled by the snapping tool (**4**).

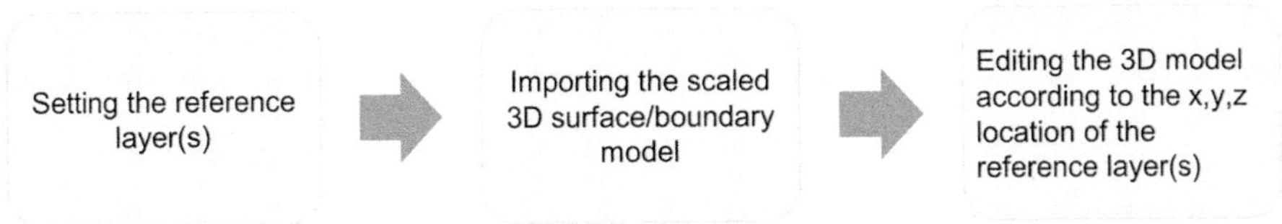

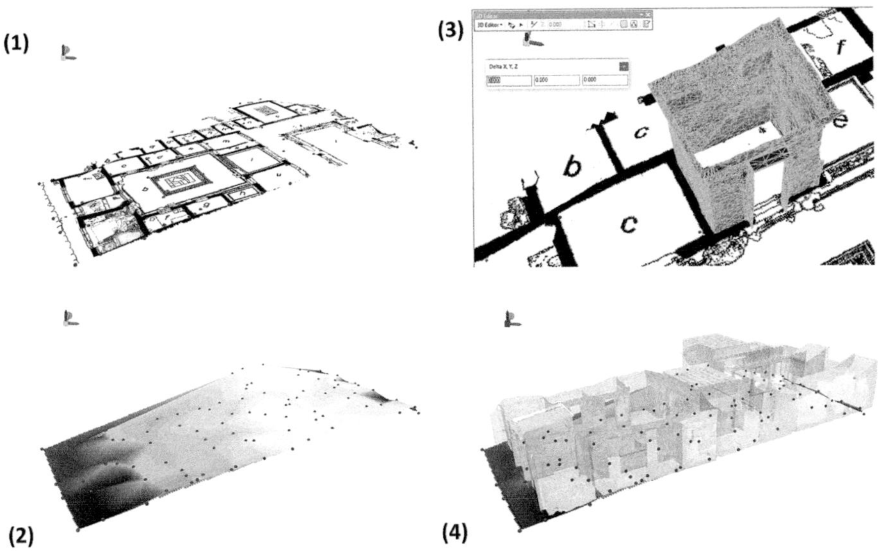

A few notes: Despite the possibility today of importing georeferenced 3D surface/boundary models into the 3D GIS using multiple other format types (see BOX 4.4), this approach is still very useful when dealing with 3D data which lacks spatial information. This is a common issue when reusing datasets from older projects. In the past, the use of laser scanning and image-based 3D Modelling was mainly adopted for working with sections and maps of sites and monuments, and for this reason ground control points were not recorded. The use of this approach permits re-contextualising and reusing archaeological records that are no longer available on site.

3.3.3 Setting up GIS basic functionalities

As a final step in the implementation of this model, topological relationships between the imported single room/environment models were checked and then each one was linked to the database information stored on the SPP website (http://www.pompejiprojektet.se/insula.php). Metadata were structured according to the architecture of the project website in order to provide users and team members of the SPP with an intuitive connection with the original documentation collected and made accessible online. According to the DB scheme, the "room" was set as the core entity. Each "room" was, in turn, connected through a relational structure with other entities, including architectural structures and documentation layers like photographs and drawings. The bespoke nature of the geodatabase allowed us to set up specific rules for maintaining data integrity and consistency. A hyperlink was added to each room's attribute table to connect each described object to the corresponding information presented on the SPP website.

The implementation of the house of CI in the form of spatially connected single room entities allowed us to define an efficient workflow of data implementation for exploring, measuring, and analysing aspects related to the material nature of the house in relation to its spatial location and the social dimension of its original space. The possibility of managing this dataset in a GDBMS allowed researchers to rethink strategies of field documentation and to start experimenting with innovative pipelines for producing archaeological documentation in situ, by taking advantage of fully 3D editing tools. Such a work pipeline can also allow researchers, and specialists of Cultural Heritage management, to contextualise and validate observations done during fieldwork with a great degree of accuracy. It also encourages them to build new interpretations based on a mutual exchange of information among scientists working in the same area, in a direct visual relationship with the object being documented (Dell'Unto et al., 2015; Landeschi et al., 2015).

3.3.4 Visualising models in the 3D GIS

As noted in Dell'Unto et al. (2015), several operational tasks can be performed by visualising and managing surface/boundary models in a fully 3D space. One of the most remarkable aspects of this is the possibility of handling spatial information diachronically by implementing different types of datasets recorded in different field seasons. These datasets not only include 3D models related to the current state of preservation of the Pompeian buildings, but also digitally scanned plans and sections that document field interpretations made by SPP team members during previous years, or interpretative 3D models providing possible views of the structures as they appeared back in 79 AD. Notably, such a visualisation can integrate and provide a holistic view of the relationships between interpretative perspective drawings and the actual geometrical references provided by the boundary models. As an additional point, there are 3D editing tools that allow users to digitise shapefiles defined by three separate x, y, z coordinate values, so as to define 3D layers that can be used to define interpretative layers corresponding to the units/features observed in the stratigraphy. 3D editing tools can also be used to analyse relationships between architectural elements in a way that cannot be observed in the field. As an example, the digitised boundaries of an alcove in the house of Caecilius Iucundus could be easily reprojected onto the wall surface of the neighbouring room (triclinium), by moving the digitised boundaries according to x, y, z coordinates and then measuring its height and width so as to assess the impact the former door entrance would have had on the triclinium (Dell'Unto et al. 2015, see below). More notably, managing 3D models in a GDBMS can trigger unprecedented opportunities for linking different types of material evidence and identifying new archaeological patterns (this is something that will be further examined in the following chapters).

BOX 3.2 Moving a 3D layer through the Delta function

The methodology described in this box can be developed using either ESRI ArcGIS Desktop or PRO; the 3D models imported in the GIS were acquired using Laser scanning; the point clouds were post-processed using Meshlab for generating the mesh and Blender for projecting the textures.

By using the "3D editing tool", users can move any shapefile to a specific location by taking advantage of several tools. By using the Delta function it is possible to shift features according to specified x, y, z coordinates. In the house of Caecilius Iucundus, the boundaries of an alcove located in Room P, West wall (cubiculum), were digitised to be moved perpendicularly to the opposite face of the wall, in order to geometrically analyse the impact that an entrance would have had on that side of the neighbouring room (triclinium), by measuring its height and width.

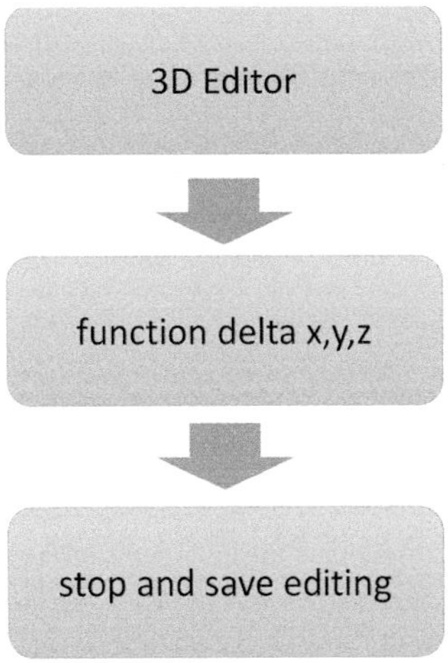

Steps for using the Delta function: (**1**) starting from the shapefile, the feature that needs to be moved is selected; (**2**) through the Delta tool the user can fill in the distance values by which the selected features will be moved along any of x, y, z planes; (**3–4**) once moved, the selected feature will appear in the new location, which in this case is the opposite wall face belonging to the neighbouring room, as we want to assess the impact the former entrance door had on that wall. To do that, a measuring tool can be used to evaluate both its original height and width.

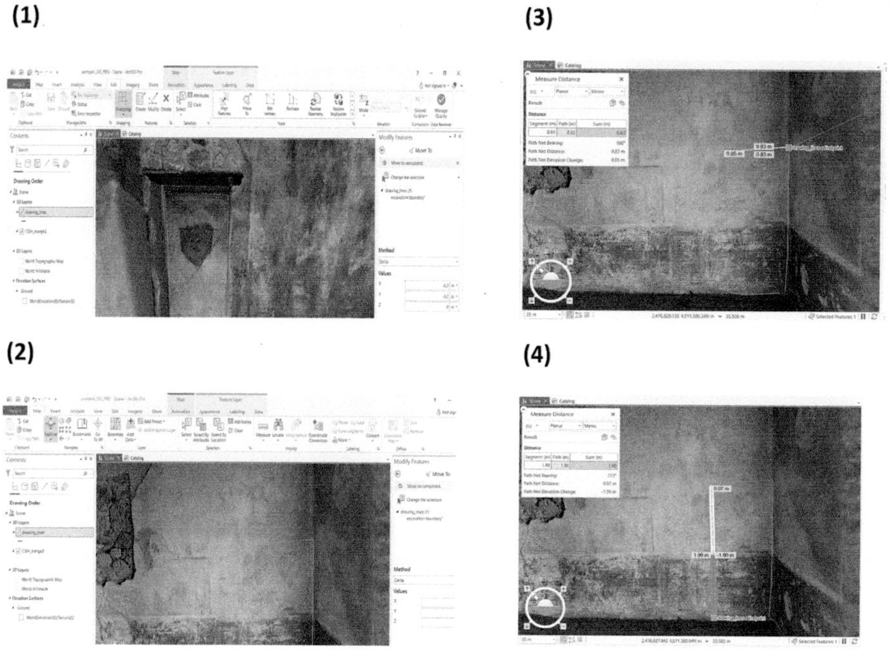

A few notes: When applying the Delta function, the current location of the selected feature is considered as the origin (0,0,0) and the feature moves according to the specified coordinates.

3.4 Towards 3D spatial analysis

When it comes to spatial data and GIS, an important aspect to bear in mind is the exploratory value of the examined datasets. Once the 3D models are imported into a GIS environment, it is possible for archaeologists and specialists to take advantage of dedicated geoprocessing tools for extracting additional information from the archaeological record. Performing advanced spatial analysis on 3D data is now an achievable goal, and this new direction perfectly aligns with the more general developments in archaeological spatial analysis over the last 40 years, such as visibility studies.

3.4.1 Archaeological spatial analysis: a quick overview

In the last 40 years archaeological spatial analysis has been much discussed and theorised, with a recently published volume (Gillings et al. 2020) presenting and discussing current trends and future directions for the application of spatial analytic methods in the discipline. It is therefore out of the scope of this book to provide an exhaustive insight into the theoretical discussion about the rise and fall (and new rise again!) of the spatially informed analyses of archaeological phenomena. This section summarises some of the key concepts that characterise this long discussion, pointing out some of the most contentious issues that still need to be addressed. What is notably absent from the current debate is a definition of a 3D-oriented approach to archaeological spatial analysis. Although part of the reason for this has been the ongoing lack of technology, it is also true that a significant limitation is posed by a still-2D-oriented attitude towards data representation and understanding of spatial phenomena that actually took place in a 3D space.

Interestingly, the concept of spatial analysis applied to archaeological studies goes back in time as far as the end of the 19th century, with the first studies conducted on-point patterns examined by anthropogeographers (Hodder and Orton 1976; Clarke and Clarke 1977; Rood 1982:26). After World War II, the rise of processual archaeology (also known as "New Archaeology") made it clear that the use of quantitative and inferential methods would have had a major impact on the discipline, leading to the development of new ways of representing and interpreting archaeological spatial information. In this sense, as several authors pointed out, "spatial analysis may be seen as a process of searching for theoretically meaningful patterns in spatial data" (Kintigh and Ammermans 1982:31).

3.4.2 Evolution of methods

The spatial analytic approach to the study of archaeological material culture heavily relied on the theoretical and methodological strands introduced by Human Geography during the 1970s (Haggett 1977), where spatial patterning was studied with statistical methods (nearest neighbour, multivariate, regression-based analysis, etc.) in an attempt to find any possible correlation between spatial distributions of artefacts, single sites, and past human activity. These correlations were thought to be caused by physiographic characteristics of the landscape or other phenomena that were suspected to have an influence on human behaviours. This is particularly true for predictive or site location modelling, developed as a method for detecting areas of past human activity based on the evaluation of predictive factors capable of influencing settlement choices made by past human groups (Kvamme 1990; Brandt et al. 1992; Warren and Asch 2000; Van Leusen et al. 2005). These predictors often included environmental elements such as distance to water sources, topography (slope, aspect), or geology, just to name a few.

In this respect, part of the criticism raised against quantitative archaeology was the exclusion of any cultural factors from the analysis of these spatial patterns and, more remarkably, the lack of "people" from the landscape that processualists archaeologists attempted to investigate (Thomas 1993; Fleming 2006). As a consequence, the use of quantitative and GIS-based methods took a different path, and during the 1990s spatial analysis was devoted to the study of human-centred forms of investigation of the landscape, where perceptual aspects of the relationship between humans and landscape such as sight, hearing, and movement were examined (Madry and Crumley 1990; Gaffney and Stancic 1991; Wheatley 1995; Llobera 1996; van Leusen 1999). Such a turn was undoubtedly affected by the theoretical debate surrounding phenomenology and experiential approaches to the study of past landscapes (Ingold 1993; Tilley 2008).

3.4.3 New directions

To sum up, as Verhagen recently discussed (2018), spatial analytic methods in archaeology can be grouped into three major categories: location modelling, movement analysis, and visibility studies. Interestingly enough, GIS is still at the core of most of these applications, and the real question is whether this technology can contribute to an advance in spatial thinking and affect the way we engage with the study of the past from a spatial perspective. An important aspect of this is discussed in Lock and Pouncett (2017), namely the "crisis of representation" that seems to characterise GIS-based technology and that prevents it from providing an adequate picture of the reality of a landscape. This problem, already pointed out by other authors (Llobera 2012), is due to the limitations in terms of data models being used in most GIS applications, which are often still characterised by the typical distinction between raster and vector data, the latter further divided into point, line, and polygon geometric primitives (see discussion Chapter 1). Such limitations might be effectively overcome if we started to think about GIS not just as a "closed box" containing all the tools needed to represent and analyse a landscape or an object of the past, but instead as an interoperable technology capable of integrating information related to different types of natural and anthropic phenomena.

Significant advances in the field of computer graphics, unmanned aerial/underwater vehicles, eye-tracking, and motion capture technology, just to name a few (see Chapter 8), have created unprecedented opportunities for capturing material and non-material aspects of reality that have never been grasped before, and have provided researchers with the possibility to experiment with new approaches in which both raster and vector information are included (see BOXs 6.1 and 6.2). Such technical achievements, however, create new problems of data representation within a GIS environment (see Chapter 8). As an example, a visual experience of a place can be described or represented in a wide range of ways. At the very basic level, computational methods allow users to establish the "connectivity" between an observer and a target point, taking into account their spatial location and the presence of any possible obstacle that could prevent their direct connection. On a more advanced level though, this information is not sufficient for qualitatively describing the visual experience of an individual, as there are additional parameters that affect visual acuity that can inform us about the actual capability of an individual to correctly perceive an image. In this sense, elements such as distance from the target or illumination conditions should be taken into account in any simulation process related to visual analysis. To cope with the complexity of a "real-world" experience it is therefore necessary to integrate multiple analytic tools, and this means using GIS as a sort of terminal for the integration of phenomena as they have been recorded in external analytical environments.

A similar example can be extended to geophysics, where multiple instruments of site investigation can now provide a holistic view of material and immaterial traces of past human activity. Although geophysics is not a new topic per se in landscape archaeology, it is now possible to three-dimensionally analyse surface and sub-surface elements and keep the visible archaeological features in spatial connection with the non-visible features (represented by recorded anomalies). What is the amount of space filled by archaeological deposits in a stratigraphy? What was the impact of historical flooding on a certain site? What level of erosion is associated with the buried structures of a Roman city? These are some of the possible questions one may ask in the phase of planning a field excavation or survey and that demand adequate instruments of data representation and analysis to be properly addressed. Models of density, accumulation, and clustering can be built upon the collected volumetric datasets and can improve the reliability of the interpretative conclusions drawn from observations of the data.

In the next chapters, case studies and examples of a heuristic use of 3D data information in GIS will be presented. Field excavations, ancient buildings and landscapes are investigated and new pipelines examined.

3.5 Summary and conclusions

The first part of this chapter provided an overview of how 3D GIS was defined across the years for answering different research questions. Multiple types of 3D platforms employed for the management of 3D spatial data were presented and briefly compared, in order to outline potentials and pitfalls of the different solutions available today. The chapter discussed emerging interest in 3D data in archaeology and remarked on the importance of focusing on how these changes impact our practice and affect the process of knowledge production.

The second part of the chapter describes one of the first experiments developed for implementing 3D GIS for supporting research carried out as part of the SPP. This specific case was chosen because of the variety of spatial data produced since the project started, and the large number of 3D models available after the laser scanning acquisition campaign undertaken in 2011. The chapter contained a few informative BOXes which illustrate the methodology initially developed for implementing 3D boundary models in the 3D GIS. This approach is still today useful when dealing with datasets which lack spatial referencing (such as ground control points), and it can be useful for reusing and contextualising datasets representing records no longer available in the physical world.

The third part of the chapter discusses spatial analysis and GIS, reviewing the way different methods evolved across the years. This part introduces the upcoming chapters by providing a theoretical background before moving on to the individual case studies.

It is difficult to foresee how these platforms will be developed in the future. The major challenge ahead of us seems to relate more to how 3D GIS will impact our way of practicing archaeology rather than what platform/s will be adopted. In Scandinavian archaeology, 3D models are used more and more as primary data acquisition sources. For this reason, it is important to find solutions for including these data within the rest of the documentation produced during the investigation. The current trend points to the development of platforms capable of including and managing multiple types of 3D datasets in the same space (3D visualisation and geodatabase), providing researchers with the opportunity to review (also in the same space) archaeological records from multiple perspectives.

4 Deploying 3D GIS at the Trowel's edge

4.1 Introduction

In the late 1990s, several experiments were carried out to test the use of 3D recording technology in support of archaeological field practice (Barceló et al. 2003; Losier et al. 2007). Laser scanning was introduced as part of several field investigations and the results were often visualised in GIS so as to produce a more comprehensive documentation of the investigation process (Doneus and Neubauer 2005; Lambers et al. 2007; McPherron et al. 2009). The possibility of using active sensors (laser scanning) for recording archaeological contexts and features retrieved during the investigation in high resolution was something new. 3D representations were no longer associated only with 3D reconstructions, but started being considered a new and sophisticated form of archaeological record, with substantial implications for the process of knowledge production.

As previously introduced in Chapter 3 (and further discuss in Chapter 7), voxel models were also tested for generating 3D, diachronic representations of archaeological deposits encountered during investigations (Bezzi et al. 2006; Lieberwirth 2008; Merlo 2016). Although the results were promising, the extended data post-processing and the lack of adequate 3D visualisation systems prevented the integration and the spread of such methods within more common excavation routines. As a result, 3D representations (such as boundary and voxel models) were used primarily during post-excavation activities for revising the operations carried out in the field and clarifying the spatial relations between the different deposits encountered during the excavation.

A clear turning point came with the spread of 3D recording and visualisation techniques based on image-based 3D Modelling. The introduction of this documentation method triggered an intense experimentation phase, in which 3D models were generated and used in the field as a primary source of documentation. Image-based 3D Modelling relied on more affordable equipment and an efficient, semi-automatic data processing workflow.

This approach allows archaeologists to create geometrically accurate 3D textured models and 3D dense point clouds from a set of digital images, which can be further processed for generating 3D boundary and voxel models. In general terms, the software calculates the camera parameters for each picture, detects and matches the geometrical features visible for each pair of successive images, and estimates the pictures' corresponding positions in space. Once the pictures'position has been estimated, dense stereo reconstruction algorithms are used for generating a geometrically accurate 3D model of the scene. The dense cloud can be exported or further processed to transform it into

DOI: 10.4324/9781003034131-5

a 3D high-resolution textured model; it can be scaled, georeferenced, and later used to extrapolate additional geometrical information (Scharstein and Szeliski 2002; Seitz et al. 2006; Callieri et al. 2011; Verhoeven 2011; Dell'Unto 2014). Today these techniques can be applied on standard desktop machines, and for more demanding computational operations, network or cloud processing services are available on demand.

Early in the history of their use, the time necessary for computing 3D models with this approach was considered one of the main obstacles for the use of this method in the field (Forte et al. 2012). Due to the high computational demand, the first experiments conducted for supporting archaeological field practice were often developed using remote online platforms. Although not ideal, these solutions permitted testing this approach in archaeological excavation environments, thus providing a general idea of the affordances of these records in relation to field investigation. Among the different platforms available at that time, one of the most popular was Arc3D, a free web-based 3D reconstruction service tailored to cultural heritage practitioners' work. This system allowed users to upload images of a scene on a server connected to a computer cluster that generated the model. Once completed, the model(s) could be downloaded from an FTP server and visualised using different 3D visualisation systems (Vergauwen and Van Gool 2006). The later spread of desktop solutions for generating 3D datasets (e.g. 3D textured models, dense point clouds) on a local computer encouraged the development of new experiments for exploring the use of this approach in multiple investigation environments.

4.2 Integrating 3D models and excavation practice: a steady progress

One of the first multidimensional digital platforms designed for supporting the work of archaeologists conducting field investigation was REVEAL (Reconstruction and Exploratory Visualization: Engineering meets ArchaeoLogy), a platform developed as a result of a collaboration between the Institute for the Visualization of History at Brown University and the Department of Electrical and Computer Engineering, at the University of North Carolina (Gay et al. 2010). REVEAL provides archaeologists with a complete data management tool designed to integrate interactive 3D visualisations and keep the archaeological documentation process entirely digital (Sanders 2011:90). Although it was very advanced for its time, the use of this product did imply a significant change to how archaeological datasets were recorded, implemented, and used by the excavation team. This was probably one of the reasons why this platform did not have a significant spread among practitioners. Archaeological practice has a strong influence on the process of data interpretation, and its transformation often comes after a long experimentation period (Taylor and Dell'Unto 2021).

In 2013, a research paper coordinated by the University of Ghent, Belgium demonstrated the effectiveness and sustainability of image-based 3D Modelling for the 3D recording and visualisation of archaeological finds (De Reu et al. 2014). Despite not addressing 3D GIS directly, this work was one of the first experiments that demonstrated the effectiveness of using 3D models in support of archaeological practice, proposing concrete solutions for integrating data derived from 3D surface models within more "traditional" excavation routines. This method is still very widespread today and has encouraged important reflections concerning the impact that 3D technology is having in archaeological field practice (Morgan and Wright 2018; Taylor et al. 2018) (Figure 4.1).

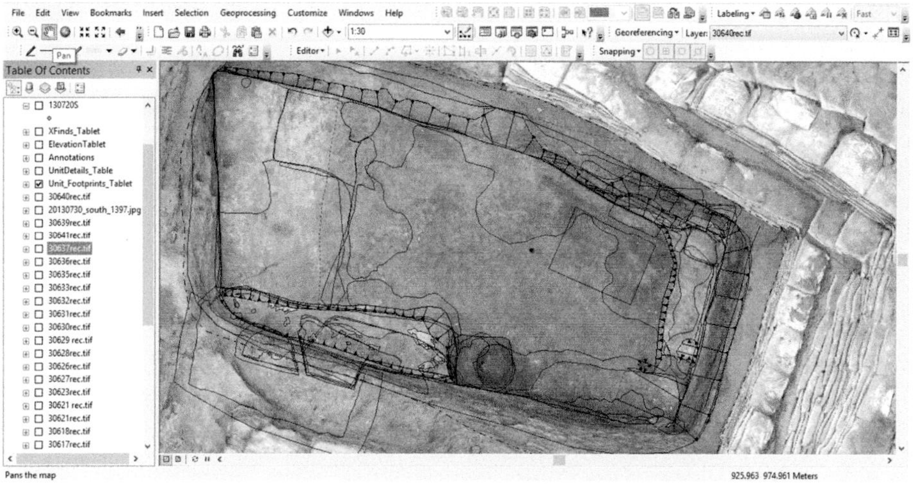

Figure 4.1 An example from Çatalhöyük; a 2D orthophoto generated from a 3D model, georectified and imported into the project intrasite GIS for planning (note the emulation of "Museum of London Archaeology Service-style" planning conventions in the GIS symbology) (from Taylor et al. 2018).

A different approach was adopted as part of the archaeological excavations at Uppåkra, Sweden where image-based 3D Modelling techniques were employed for recording the field investigation process of a late Neolithic grave detected in 2010 using ground-penetrating radar (GPR) (Callieri et al. 2011; Larsson et al. 2015). Once created, the 3D models were spatially aligned and visualised using Meshlab, an open-source system for processing, editing, and visualising 3D triangular meshes (Cignoni et al. 2008). This method could visualise multiple 3D models in the same digital space, each one representing a different stage of the field investigation. The system was used by the archaeologists working in the field for revising the spatial relations among the different contexts identified during the excavation (Dellepiane et al. 2013). This experiment demonstrated the advantage of having high-resolution 3D representations of archaeological contexts and features available during the field interpretation. Although very powerful, Meshlab is not designed for the management and analysis of archaeological records and does not possess the same capabilities as a GIS for handling a large variety of spatial data.

In 2012, Opitz and Nowlin published a paper describing a methodology for importing photogrammetrically derived meshes into a GIS. This approach allowed them to use the 3D models in spatial relation with the archaeological datasets included in the excavation geodatabase (Opitz and Nowlin 2012), thus demonstrating the possibility of combining 3D surface models and 3D GIS at the trowel's edge. Since then, several experiments have taken place for assessing the impact of this new approach in support of archaeological practice, providing the community of practitioners with enough of a background to begin a critical evaluation of the use of these technologies in the field (Berggren et al. 2015; Huggett 2017; Morgan and Wright 2018; Taylor et al. 2018; Jensen 2018). As recently pointed out, in the last decade 3D visualisation at the trowel's edge has received a lot of attention, and it has been argued that more focus should be directed towards the use (or reuse) of these datasets during post-excavation activities (Katsianis et al. 2021:1).

In the following chapters several examples will be presented and discussed where 3D GIS was employed for supporting post-excavation work, but it is important to underline that in order to use 3D models as primary data sources during post-excavation activities, it is necessary to establish robust field acquisition and validation routines in which the data captured can be compared against the original materials encountered during the investigation.

In the coming pages, various field projects will be presented so as to provide an overview of some of the possibilities offered today by 3D GIS in support of archaeological field practice.

4.2.1 Çatalhöyük

Among the different projects that experimented with the use of 3D GIS platforms in support of archaeological field practice, the Çatalhöyük research project (http://www.catalhoyuk.com/) remains an important example. Çatalhöyük is an archaeological site located near the town of Çumra in the Konya region of Turkey. The site was first excavated between 1961 and 1965 by James Mellaart (Hodder 2000) and is considered one of the key sites for understanding human prehistory.

This project has a long history of engaging with digital archaeology and 3D models (Tringham and Stevanović 2012), and since 2009, several experiments have been carried out for mapping potentials and limitations of 3D recording technology in support of field practice (Forte et al. 2012; Berggren et al. 2015; Dell'Unto 2020). Initial experiments were part of the 3D Digging project, where different visualisation platforms were used for supporting the work of field archaeologists and specialists operating on-site (Knüsel et al. 2013; Campiani et al. 2019). After a long experimentation period, in 2013 the project decided to include image-based 3D Modelling as an additional method for recording spaces, buildings, and features encountered during the investigation. In order to include the 3D models produced with this technique in the project database, several protocols and workflows were established (Taylor et al. 2018) and within a short period, an extensive library of 3D models was available. The possibility of combining 3D surface models with the rest of the graphic documentation encouraged different forms of interaction with the data collected during the investigation, providing archaeologists and specialists with a broad and more accurate perception of the archaeological records (Carpentier and Lundin 2015). The photographic material produced on-site was processed using Agisoft Photoscan (today Metashape). Once created, the 3D models were georeferenced, exported as collada files and imported into an ESRI Geodatabase Management System (GDBS), to be visualised in GIS along with the rest of the graphic documentation produced on-site.

The GIS adopted by the project was ESRI ArcGIS, one of the most widespread platforms for managing 3D georeferenced data. Although older versions of the software did not allow users to import high-resolution meshes, the 3D models produced within the project were nevertheless of sufficient quality for supporting field documentation and analysis. Later versions of the software have overcome these issues, and today high-resolution mesh models can be quickly implemented and visualised through ArcGIS PRO (Figure 3.6), a new 3D GIS platform oriented towards 3D Data (see BOX 4.4). Despite these initial limitations, the possibility of using 3D GIS during the field investigation promoted a more reflexive approach (Berggren et al., 2015), providing the excavation team with a broader and more resolute overview of the ongoing field investigation and with a new set of possibilities for testing new approaches.

4.2.1.1 Reviewing relations on a large scale

The implementation of such methods triggered the development of a large library of 3D models, representing features and structures encountered during the field investigation. These data were used on multiple occasions throughout the various excavation seasons for reviewing the different steps undertaken during the excavation. In most cases, the archaeologists used the models to observe (in 3D and chronological order) the sequence of contexts and features retrieved in the field across different seasons, thus gaining a detailed understanding of the spatial relationships between structures which were no longer visible on-site (Figure 4.2).

On a few occasions, the 3D models were used in the 3D GIS by various team members in the form of point clouds so as to gain a detailed visualisation of the micromorphological characteristics of surfaces such as floors or platforms encountered during the excavation (see Lercari and Lingle 2015). This approach was particularly effective for the detection of contexts only visible under special light conditions (e.g. burials or postholes) or for carrying conservation work (Lercari and Lingle, 2015:246). The methodology involved creating a 2.5D raster image, derived from high-resolution 3D point clouds, which emphasises the rate of maximum change in Z value from each cell (Slope) (ESRI 2020). The possibility of detecting surface anomalies at the trowel's edge can have a strong impact on the excavation strategy, providing important indications about how to progress with the field investigation (BOX 4.1).

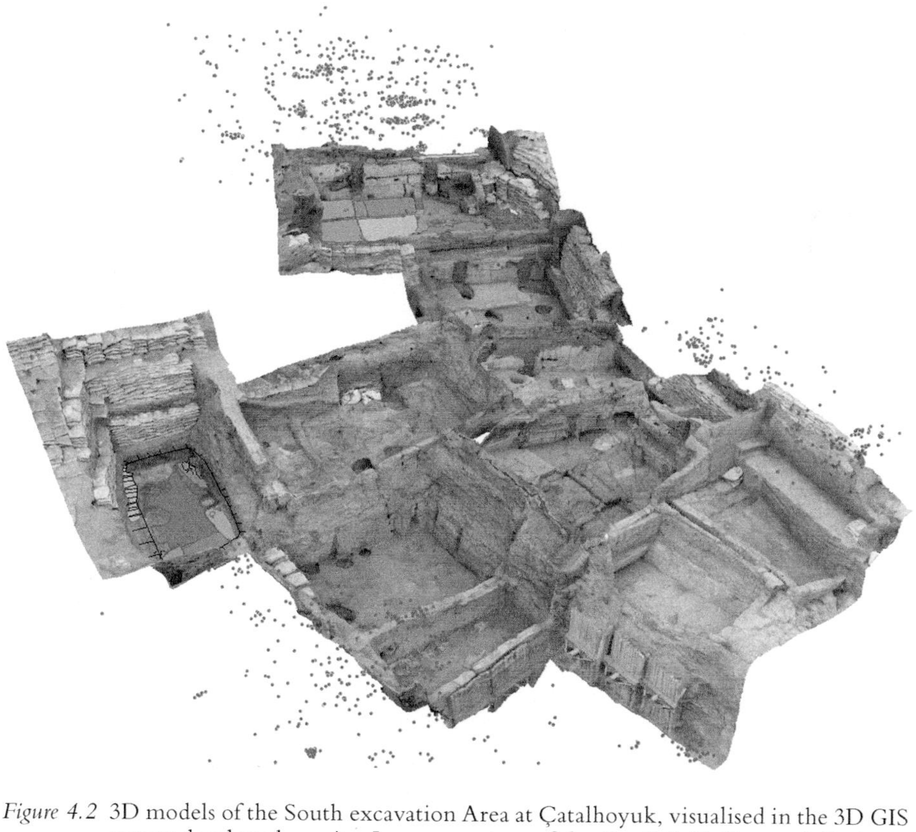

Figure 4.2 3D models of the South excavation Area at Çatalhoyuk, visualised in the 3D GIS system developed on-site. Image courtesy of the Çatalhöyük Research Project.

BOX 4.1 Detecting anomalies at the trowel's edge

The methodology described in this box can be developed using either ESRI ArcGIS Desktop or PRO; the 3D models imported in the GIS were acquired at the archaeological site of Uppåkra (southern Sweden) using a DSLR digital camera; the images were processed using Metashape 1.7. ArcGIS PRO was used for the visualisation and analysis.

This BOX provides a summary of the steps needed to generate a slope visualisation of the excavation surface, used for performing micromorphological analysis at the trowel's edge.

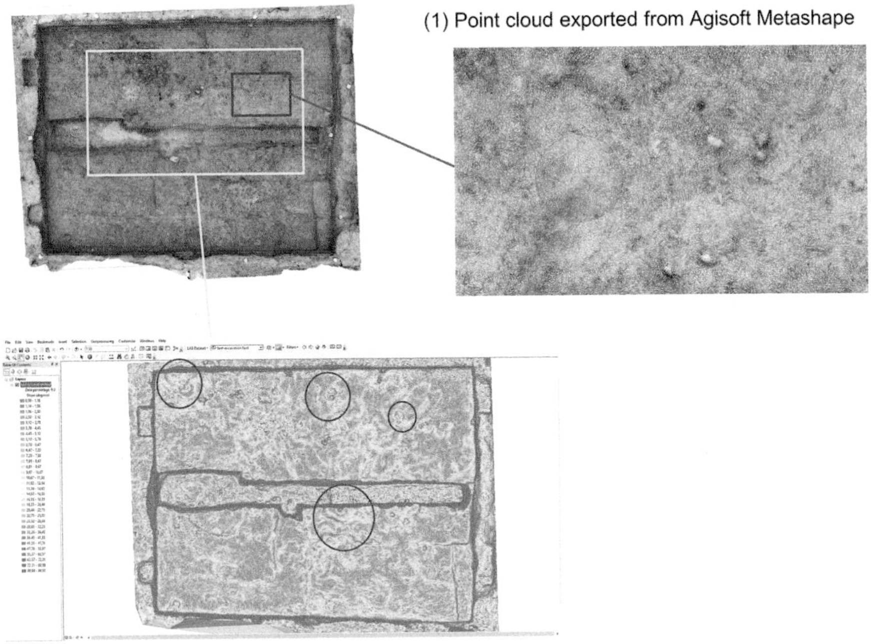

(1) Point cloud exported from Agisoft Metashape

(2) Points cloud visualized in GIS as slope

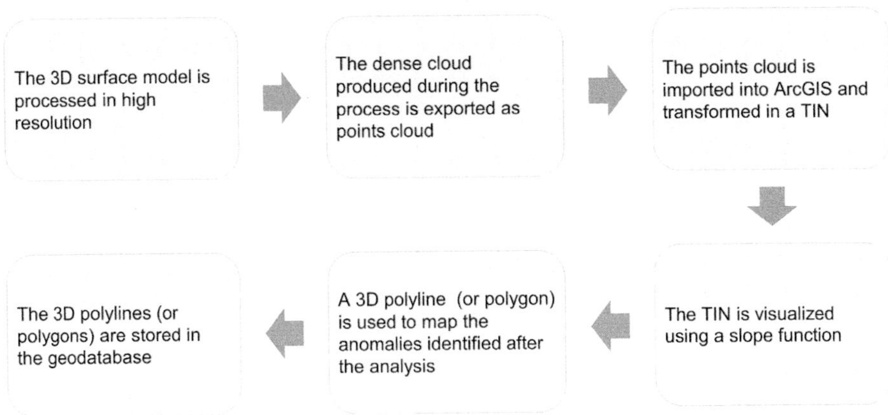

This method is very diffuse today and it consists of (a) image-based 3D Modelling for generating a dense point cloud of the horizontal excavation surfaces, (b) a GIS platform for creating a 2.5D raster visualisation of the micromorphological characteristics of the surface, and (c) 3D polygons and polylines for recording the anomalies detected during the analysis. This approach can be employed for detecting ground anomalies at the trowel's edge such as postholes, deposits or cuts. The 3D model representing the site surface is processed in high resolution and exported as a point cloud (**1**) (e.g. LAS format). Once in the GIS, the point cloud is converted into a shapefile (multipoint) and then into a triangulated irregular network (TIN) (LAS /Multipoint/TIN). Using the symbology tool available in ArcGIS, the TIN can then be visualised as a slope and later used to identify anomalies on the ground (**2**). 3D polylines can be created in the geodatabase and used to map the anomalies visible in the slope directly onto the TIN. In order to achieve a good result, it is important to find the right visualisation settings. The parameters can be customised using the symbology tool.

A few notes: The pipeline described in this BOX can be developed using almost any GIS platform as it is based on the transformation of a 3D point cloud into a 2.5D surface (TIN). However, the use of a 3D GIS platform for visualising the results of the analysis in spatial relation with the rest of the 3D documentation (e.g. voxels, boundary, surface or 3D polylines) increases the chance of identifying new archaeological relations that might be difficult to spot in a 2D GIS environment.

4.2.1.2 Working with 3D puzzles

A significant example of the impact that 3D models had during the field investigation occurred in 2015, when a painted plastered head (F21666) was retrieved during the excavation of one of the north area buildings (Building 132). The feature was immediately removed because of its fragility, preventing the researchers from studying it in spatial relation with the rest of the materials retrieved during the investigation (Lingle et al. 2015; Dell'Unto 2018). By using the 3D models stored in the geodatabase, the archaeologists were able to identify an early digital version of Building 132 in which the feature was still in place (Klimowicz and Tung 2015).

In this older version of the building, the plastered head was covered with a layer of clay which prevented a clear reading of the relation between the artefact and the rest of the contexts. For this reason, a 3D model of the plastered head, after conservation work, was created and imported into the 3D GIS in spatial relation with the latest version of the building. Thus, an important archaeological feature could be recontextualised and support the work of the archaeologists working in the field. F21666 was interpreted as part of a larger feature visible in the southwest corner of the building, suggesting a symbolic relationship between the feature and the rest of the structure (Lingle et al. 2015:276). This result would have been difficult to achieve without the support of a 3D visualisation environment capable of linking various datasets collected across time and space.

Figure 4.3 The image displays the plastered head (F21666) after conservation implemented in ArcGIS PRO in spatial relation with Building 132. Photographic acquisition for producing the models: Jason Quinlan, Marta Perlinska and Nicolò Dell'Unto; 3D models: Nicolò Dell'Unto. Image courtesy of the Çatalhöyük Research Project.

BOX 4.2 Aligning objects in the 3D GIS

The methodology described in this box is developed using ESRI ArcGIS Desktop/PRO; the 3D models imported in the GIS were acquired using a DSLR digital camera; the images were processed using Metashape 1.7. and Meshlab. For the alignment of the different models ArcGIS PRO was used for the visualisation and analysis.

This BOX provides a detailed explanation of the experiment described in subchapter 4.2.1.2 The following pipeline can be easily re-adapted for solving similar situations encountered in the field. Once identified, the 3D model of Building 132, displaying the plastered head (F21666) in its original position before removal, was re-processed in Agisoft Photoscan (today Metashape) to improve geometrical resolution and texture quality (**1**). This permitted gaining better visibility of the details observable on the model and increased the possibilities of finding geometrical matches to use for a later alignment.

After conservation work, a new version of the plastered head was created and imported into the 3D GIS in spatial relation with the rest of the 3D documentation (**2**). Due to the low number of control points for georeferencing the feature (F21666), Meshlab was used to scale and align the plastered head created after conservation using the plaster head visible in the 3D model of Building 132 as reference (**3**). Once aligned, the 3D model generated after conservation work was exported as a collada file and imported into the project geodatabase (Figure 4.3).

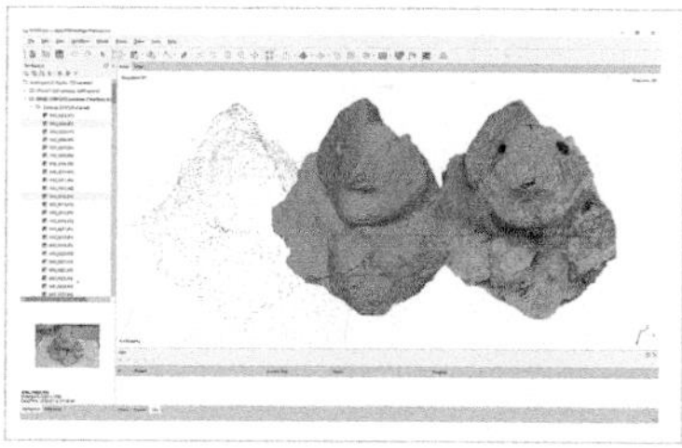

(1) Agisoft Photoscan

(3) Model Visualized in ESRI ArcGIS

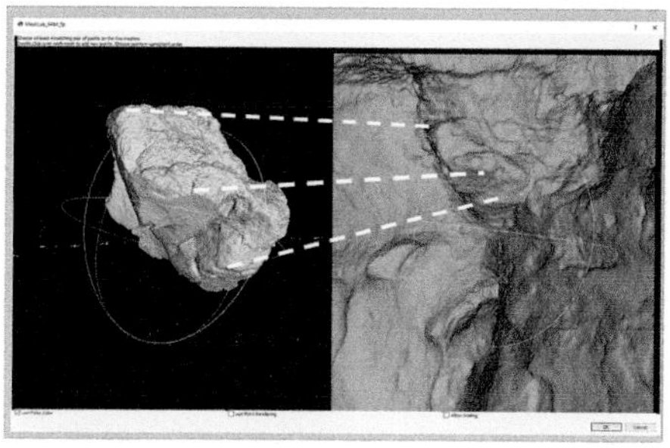

(2) Alignment in Meshlab

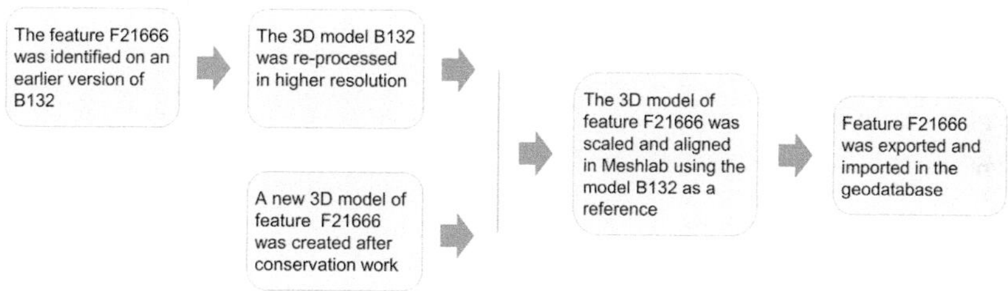

A few notes: When this pipeline was developed (2015), several tools today available in Agisoft Metashape (earlier Photoscan) and ArcGIS PRO were not yet implemented. Using more computational resources and the latest version of Agisoft Metashape, today the 3D model of Building 132 could be processed at a much higher resolution and exported in a 3D format more compatible with georeferenced 3D environments. In situations in which the features to be aligned contain more clear geometrical references, the scaling and the alignment could be performed directly in Metashape. However, considering this specific situation, Meshlab (or similar software) would, today, still represent the best solution for performing the alignment.

4.2.2 Kämpinge

The archaeological site of Kämpinge is one of the first case studies where 3D GIS was used as the main documentation management system for gaining a more comprehensive understanding of all the pieces of archaeological evidence encountered during the excavation (Dell'Unto et al. 2017). The site is located in SW Sweden and is one of the several coastal sites in the Öresund region dating from 8,500 to 6,000 cal BP (Brinch Petersen 2015). The field investigation was conducted across four seasons (2014–2017) and included a team of field archaeologists with basic training in GIS and 3D visualisation technology. After preliminary experimentation in 2014, in 2015 the team decided to use tablet PCs and portable workstations for using 3D GIS directly in the field. The possibility of accessing the entire 3D documentation at the trowel's edge encouraged discussion, intellectual engagement and allowed controlling the quality of the models in the presence of the archaeological evidence (Dell'Unto et al. 2017).

The field documentation system developed in Kämpinge relied on four key elements: (**1**) image-based 3D Modelling for generating 3D surface textured models of contexts and features encountered during the investigation; (**2**) Real-Time Kinematic (RTK) Global Positioning System (GPS) for establishing the site topography and measuring the ground control points (GCPs), used for georeferencing the 3D models; (**3**) an excavation geodatabase designed for storing the 3D records with the rest of the field documentation (3D surface and boundary models) (Figure 4.5); and (**4**) a 3D GIS platform for producing graphic and textual documentation (3D drawings and context description).

The process can be summarised as follows:

1-Once detected the contexts were cleaned and photographed for producing a 3D textured surface model that was sufficiently accurate for documenting the archaeological evidence.

2-A grid of GCPs was distributed around the trenches and measured with RTK GPS, used for georeferencing the models. The establishment of a fixed grid allowed reducing errors in scale and projection, making the entire process faster.

3-The 3D models were exported and implemented in the 3D GIS directly in the trench. This operation was carried out using a tablet PC Microsoft Surface 3 Pro equipped with 8 GB of memory RAM, a 64-bit operating system, and an Intel HD Graphics 5000. For larger computational operations we used a workstation Lenovo ThinkPad W540, Intel® Core™ i7–4900MQ 2.8 GHz, 32 GB RAM, Nvidia Quadro K2100M 2 GB DDR5.

4-The 3D models were imported into the excavation geodatabase and visualised in ESRI ArcScene as geometrical references for producing the graphic documentation. These operations were carried out in the field, and at the time of the excavation (see BOX 4.4).

BOX 4.3 Setting up the field reference system

The methodology described in this box is developed using Total Station or RTK GPS for recording the ground control points (GCPs) established during the field investigation. These are used as spatial references for georeferencing the 3D models generated during the excavation. ESRI ArcGIS Desktop/PRO is used to visualise the GCPs; the 3D models were processed using Metashape 1.7.

This BOX illustrates the process of setting up a field reference system for spatially locating and implementing 3D models in a GDBMS during the field investigation process.

Before starting the field investigation, a reference system of at least four GCPs are established for scaling the 3D models and defining their actual location according to real-world coordinates. The way GCPs are distributed across the site strongly depends on the type of excavation approach undertaken during the investigation, and on the geometrical characteristics of the contexts and features encountered during the investigation (see BOX 4.4, 4.5 and 4.7).

In order to reduce the use of Total Station or RTK GPS and minimise their time of usage in the field, a possible strategy is to set a large number of ground control points around each archaeological trench (1 GCP every 1.5 meters), (**1**) even including areas of possible expansion.

Markers used as GCPs should be made of a waterproof (ideally not shiny) material and should be cleaned before acquiring the set of images for producing the model. Once measured, the 3D points are exported as ASCII files and imported into the geodatabase (**2**). They can be used in the 3D GIS for checking the quality of the projection of the 3D models (once imported into the geodatabase), for checking the measured position of any object recorded in the field and for comparing them against the reference basemap used in the project (**3**). The GCPs must be visible in the photographs taken for generating the model, should be photographed from multiple perspectives and should be treated as integral parts of the feature or context being recorded. A 3D model which does not include ground control points (or which has an insufficient number of GCPs) will be very hard to use in a field investigation. If this is the case, a possible workaround is scaling the 3D model and then manually georeferencing it in GIS by using the 3D editing tool, but for this to work it is still necessary to have at least one recognisable segment in the model for which the position and size is known (see BOX 3.1).

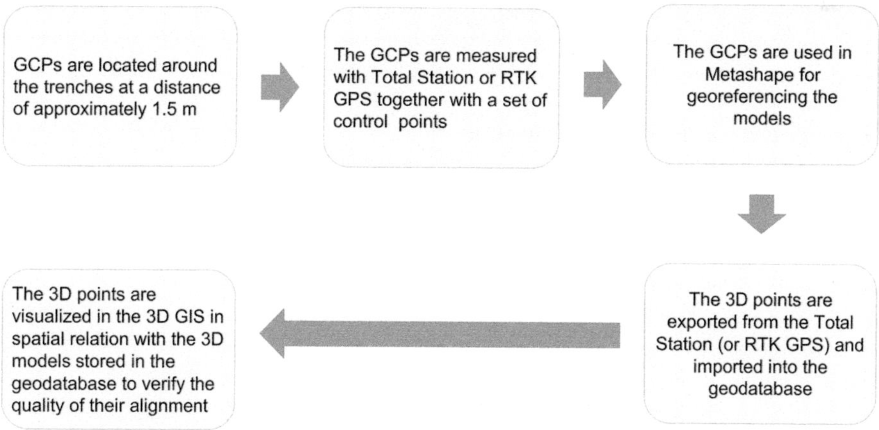

A few notes: In the situation in which GCPs need to be constantly updated (e.g. when working in open area excavation or when using large trenches), we recommend setting a system of control points (not necessarily related to the excavation

(*Continued*)

topography) for checking the quality of the new GCPs imported into the system. Each time a new group of ground control points is recorded, the grid of control points not related to the site topography are included in the measurement; in this way, it will be possible to verify the quality of the new GCPs included in the topographic system or (in case something goes wrong) adjust the new set of points according to the main topographic system.

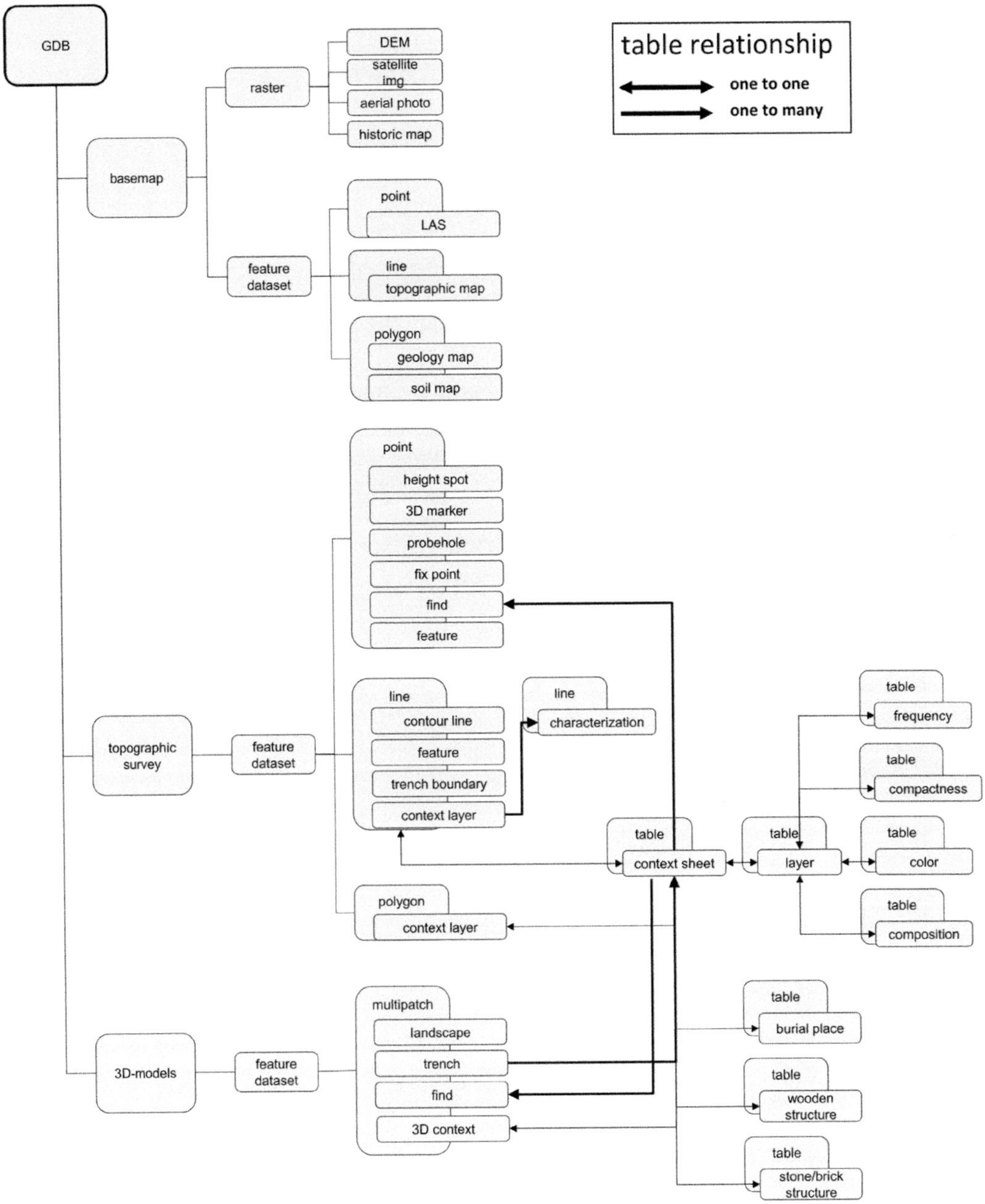

Figure 4.4 Site geodatabase scheme designed for supporting the site investigation in Kämpinge.

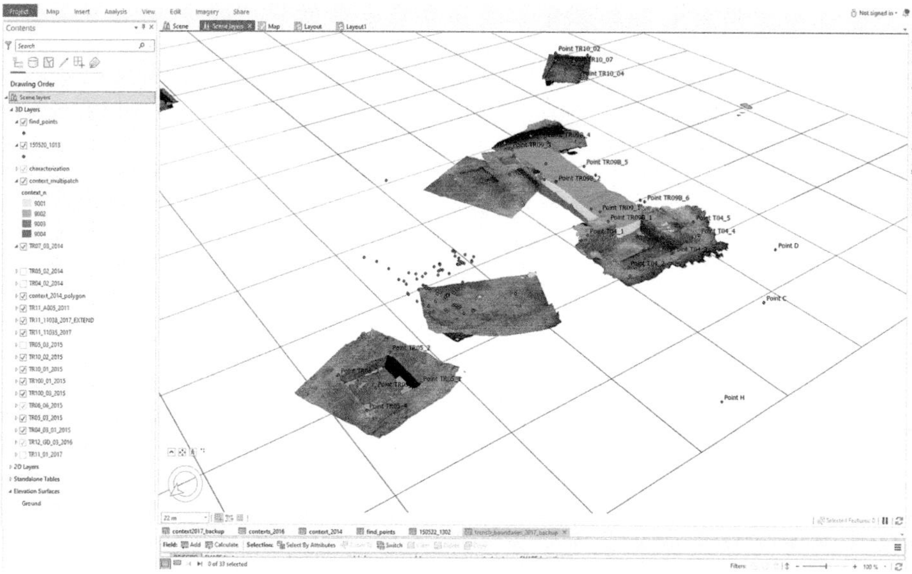

Figure 4.5 Example of the 3D GIS system implemented on-site, image courtesy of the Kämpinge project.

Across the different excavation seasons, several aspects of this approach were modified to expand and include different types of 3D datasets in the 3D GIS (mainly 3D surface models, 3D boundary representations and 3D point clouds). The constant development of new (and more efficient) versions of the software, along with the improvement of portable workstations, encouraged the team to further explore the use of 3D GIS in support of field practice. The use of the system in the field (through tablet computers or workstations) allowed gaining a thorough overview of the entire investigation process across time, providing the archaeologists with the opportunity to review and compare (in 3D) records retrieved across the different seasons.

BOX 4.4 From Image-Based 3D Modelling to 3D GIS

The methodology described in this box can be developed using ESRI ArcGIS Desktop or PRO; the 3D models imported in the GIS were acquired using a DSLR digital camera; the images were processed using Metashape 1.7. ArcGIS PRO was used for visualisation and analysis.

This BOX provides a summary of the necessary steps for importing georeferenced 3D textured surface models into an ESRI geodatabase. Once created in Agisoft Metashape the 3D surface model is georeferenced using the GCPs measured (preferably) with Total Station or RTK GPS and exported as an.obj file (**1**). Before exporting the model, it is possible to optimise textures and polygons and select the projection system used in the project (**2**). For converting the 3D model into a multipatch file, use the function "import 3D model" (in the toolbox), then

(Continued)

(**a**) select the model to be imported in the geodatabase, (**b**) provide indications concerning the projection system, and (**c**) indicate the location where the multipatch file should be saved in the geodatabase. Once converted into a multipatch file, the 3D surface model is associated with an attribute table and can be visualised in spatial relation with the rest of the documentation (**3**).

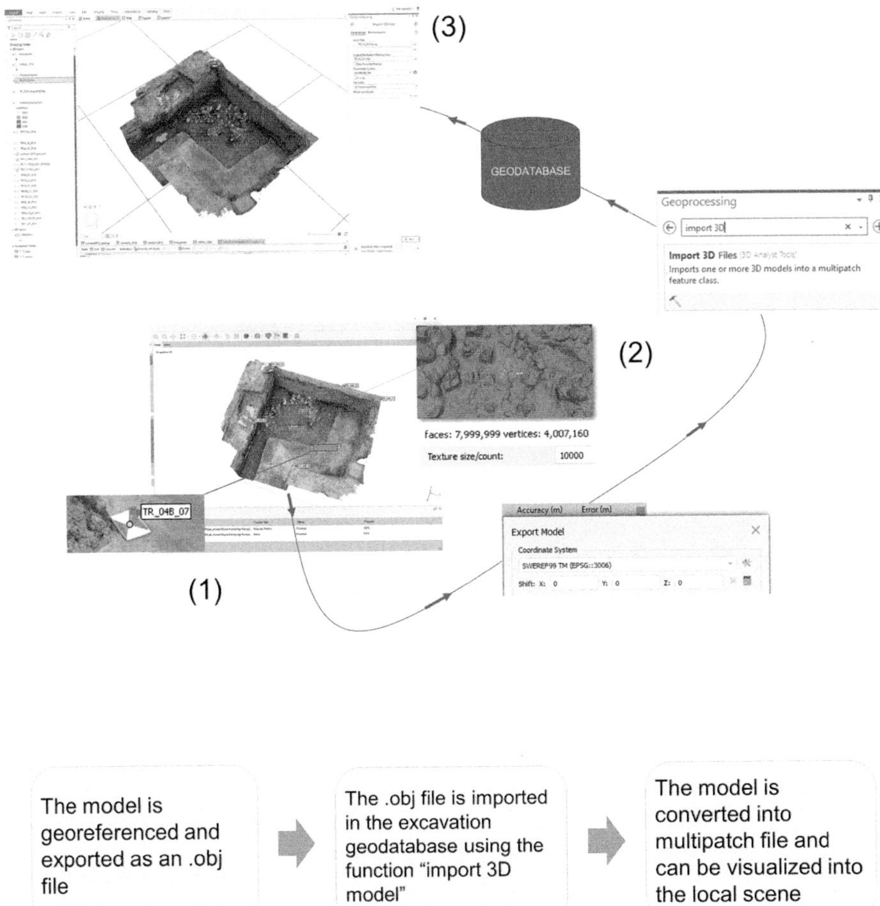

A few notes: Previous versions of ArcGIS did not allow importing.obj files. Collada or.3ds files were usually employed for carrying out this operation. However, using such formats for importing georeferenced 3D models has led, in several cases, to issues connected with scale and projection. This was solved by replacing the 3D model in the geodatabase exported from Metashape in geo coordinates with an identical version exported in local coordinates. This operation was carried out in ArcScene using the function "replace multipatch".

4.2.2.1 Recording contexts and features in 3D

Despite the possibility of producing virtually unlimited digital datasets for creating the models (digital pictures and 3D scans), before starting the excavation it is essential to set a reference system (see BOX 4.3) and define a strategy concerning the number of digital records that will be produced during the investigation. By looking at the excavation characteristics (e.g. size of trenches; type of structures; features and context expected), It is possible to establish a maximum and minimum number of pictures or scans to produce during the investigation for the description of the field evidence. This way reasonable control can be maintained over the volume of data produced during the excavation and the time needed for the processing.

Despite their efficiency in recording geometry, laser scanning techniques are considered more problematic for the representation of colour information. Colour is an important indicator in the field and is usually one of the first factors used for identifying a new deposit (Roskams 2001). Different from laser scanning, image-based 3D Modelling provides better control over colour information. The textures generated using this technique are directly derived from the photographic dataset employed for constructing the model's geometry. However, when using this method, it is important to keep in mind that despite the possibility of choosing among many different parameters for generating the texture(s), the final result mainly depends on the quality of the original dataset and on the environmental conditions in which the images were acquired.

For creating 3D models in the field, cloudy weather conditions (in which the diffuse light naturally reduces the presence of sharp shadows in the pictures) are more favourable, but it is not always possible to wait for the right light conditions and for this reason,

Figure 4.6 Use of sheet for obtaining a more homogeneous distribution of the light in the trench (Picture taken by Giacomo Landeschi).

strategies such as creating an artificial shadow over the trench can be adopted for obtaining a more homogeneous distribution of light (Figure 4.6).

From a technical perspective, these aspects could be partially overcome by using the RAW image format which provides more colour control (Kimball 2014:62); however, this would require the use of additional software in the field for the management of the RAW data; with a consequent increase in the time needed for processing the model (van Riel 2016:56). Issues associated with colour representation are usually solved by linking the 3D models in the geodatabase with textual descriptions of the context (preferably performed at the trowel's edge). In cases when 3D models are used for producing multi-context representations (as in one model showing multiple contexts), 3D polylines (or 3D polygons) can be used for highlighting the areas where the contexts are visible in the model. In cases when a single model is used for recording a single context, the colour information can be described directly in the attribute table associated with the model.

Before starting the field investigation at Kämpinge, several tests were performed for establishing the number of pictures needed and the parameters necessary for obtaining sufficiently high-resolution 3D surface models that would be processed and visualised in the field (Dell'Unto et al. 2017). During these tests we took into account factors such as (**1**) the hardware available for the processing, (**2**) the type of camera used for producing the images, (**3**) the geometrical characteristics of the objects to document, and (**4**) the availability of devices (such as extendable poles) for taking pictures from unreachable positions. Due to the kind of features expected during the investigation and the camera available, we decided to establish an average of 60 images per trench (2m × 2m). The datasets were processed and georeferenced in the field using Agisoft Photoscan, imported into the excavation geodatabase and visualised in ArcScene (see BOX 4.4). Then, field tablets were used for drawing the contexts visible on the models directly in the 3D GIS (see BOX 4.5).

BOX 4.5 Using 3D Polylines and Polygons in ArcGIS PRO

The methodology described in this box is developed using ESRI ArcGIS Desktop/PRO; the 3D models imported in the GIS were acquired using a DSLR digital camera; the images were processed using Metashape 1.7. ArcGIS PRO was used for the visualisation and analysis.

3D GIS platforms can be used for generating 3D graphic documentation of any kind of geometrical feature visible on the models. The method presented in this box can be employed for (1) using 3D polylines to outline contexts and features visible on 3D surface models, (2) drawing sections or multi-context maps and (3), even for using 3D polygons for producing volumetric representations of deposits. Once imported into the geodatabase (see BOX 4.4), the 3D models are displayed in the 3D GIS and can be used as palimpsests for creating the documentation (Figure 4.5). The first step is creating the 3D shapefile in the excavation geodatabase (polylines or polygons). If using ESRI ArcGIS this operation is performed in "ArcGIS catalogue"; when creating the shapefile be sure to select Z Value (**1**). Once in "edit mode", the shapefile can be used for drawing the features using the 3D model as a geometrical reference. The editing tools available in ArcGIS PRO can be used for gaining more control over the drawing. By using the "Symbol

Property Editor" the appearance of the individual lines can be modified so as to represent different archaeological planning conventions (**2**). It is also possible to use 3D polygons to characterise contexts visible on the surface or for constructing 3D models of the archaeological deposits. For retrieving information concerning the volume, it is necessary to transform the polygon into a multipatch file when creating a boundary model from a set of polygons (**3**).

(1)

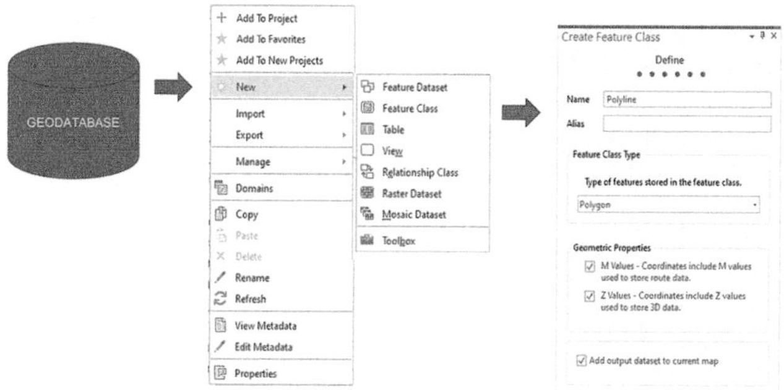

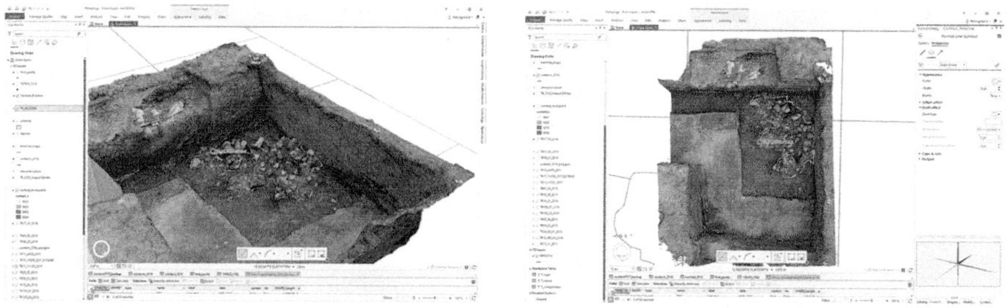

(2)

A few notes: To verify the quality of the graphic documentation, it is best to create the drawing in the field. Creating the 3D graphic documentation at the trowel's edge allows the recorder to verify the quality of the representation against the physical evidence. Features, cuts, and deposits can be documented using either polylines or polygons. Each of these have several advantages and disadvantages: the use of polylines provides a more similar result to what is usually produced by more traditional means. But polyline shapefiles can't visualise quantitative aspects of the information in the GIS. Quantitative data related to surfaces or deposits can be more easily mapped using 3D polygons. However, the use of 3D polygons for recording contexts and features visible on the 3D surface model can lead to unexpected and undesirable graphic visualisations. These are usually due to a geometrical mismatch between the shapefile (that is usually characterised by a lower number of polygons and vertices) and the 3D model used as graphic reference (which is more geometrically detailed). This problem is less evident when using the 3D polygon for creating boundary models. This type of representation is more suited for representing deposits rather than context interfaces.

(Continued)

(3)

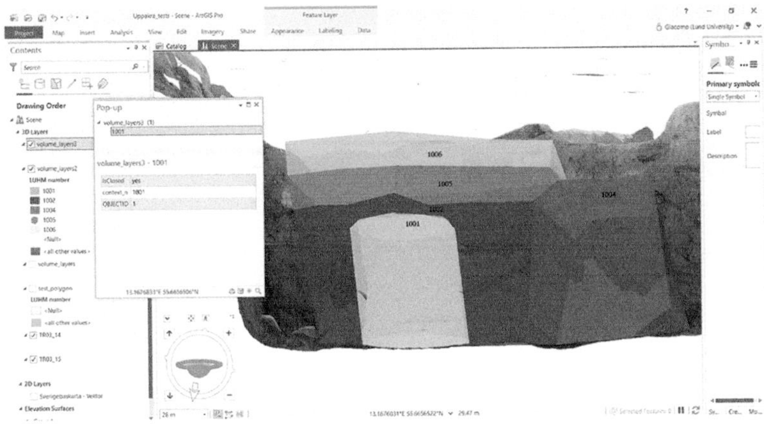

3D POLYLINES

The shapefile polyline is created in the geodatabase → The shapefile is used in edit mode for recording the feature visible on the model surface → The symbology tool is used for representing the 3D lines using different planning conventions

3D POLYGONS

The shapefile polygon is created in the geodatabase → The shapefile is used in edit mode for recording the feature visible on the model surface → Once created, the 3D polygon is converted into a multipatch file

4.2.2.2 3D graphic documentation

The use of GIS for digitising hand drawings and (later) drawing directly on orthoimages has been a practice in use since the 1990s (Eiteljorg et al. 2002). However, due to the complexity in the past of using computers in excavation trenches, this operation was usually performed during post excavation. In Scandinavian archaeology, graphic documentation is often produced by interpolating points acquired with the total station in a GIS. The surveyor uses the pole and theodolite to record the artefact distribution and outline contexts directly in the field. Each group of points is associated with a code that corresponds to a specific feature, which is visualised in the GIS according to the symbology (or planning convention) adopted during the investigation. Both approaches have advantages and disadvantages. Digitising hand drawings was more accurate, but extremely time-consuming and did not add any additional qualities to the original drawing (except its digitisation). On the other hand, using orthoimages as references for drawing contexts identified during the excavation can be a valid solution, but only recently, with the spread of more powerful computers, has it become possible to use these images on-site and thereby produce graphic documentation at the trowel's edge (Taylor et al. 2018). Total Station or RTK GPS are fundamental instruments in archaeology, and most archaeologists are familiar with their use. These instruments record the archaeological evidence with accuracy and precision, reduce the amount of time spent in the field producing the graphic documentation, and permit maintaining visual control over the features being documented. However, when it comes to drawing context boundaries, relying only on a few measured points per context (especially with a total station) does not ensure an accurate boundary description of the unit itself.

The possibility of combining and using tablet PCs and orthoimages on-site for creating the drawing at the trowel's edge represents an important improvement, as it allows archaeologists to characterise the information visible in the field in greater detail, verifying the quality of the drawing in real time. The downside of this approach is the impossibility of documenting information in 3D, which instead is done automatically when using the Total Station or RTK GPS.

3D GIS promotes the use of 3D surface models as a graphic reference for performing a drawing in 3D. This approach can be quickly implemented in the field (see BOX 4.4 and 4.5) and allows a more dynamic recording of the contexts and features visible on-site. The possibility of reviewing the drawings from multiple perspectives encourages an active engagement with the retrieved materials in the field. Current archaeological planning conventions are designed to represent three dimensionality in bidimensional representations, but the use of 3D polylines (or other type of 3D datasets for performing the drawing) poses important questions about the way graphic documentation should be performed in a 3D environment, and about whether the traditional drawing conventions should be used when operating in a non bidimensional environment (van Riel 2016).

A downside of this method is related to the outdoor use of tablet PCs, which can be difficult under specific weather conditions (direct sun or rainy weather). These issues can be improved using a matte anti-glare screen protector on the device and setting the device's brightness level to maximum. This will impact battery life but will improve the experience outdoors.

The use of 3D georeferenced surface models as palimpsests for producing 3D graphic representations of contexts and features has been discussed as part of various archaeological projects. Early versions of 3D GIS platforms possessed technical limits connected to the management of high-resolution models (e.g. ArcScene) and thus affected the quality of the 3D drawings (Verhoeven 2017). However, if on one hand, these aspects were particularly relevant when documenting large representations of landscape, they have had a more limited impact when using this approach for recording features and contexts during excavation. Nowadays 3D GIS platforms can import 3D models at much higher resolution (Figure 3.5) providing the archaeologists with the possibility of producing more resolute 3D drawings (Table 4.1).

Table 4.1 A comparison of the different techniques currently in use for performing digital drawings

Recording method	*At the trowel's edge*	*Post excavation*	*Allows creating detailed drawings*	*Allows high control*
Digitising hand drawing	**No**: even if it is technically possible to digitise hand drawings at the trowel's edge, it would be a repetition of labour and an unnecessary extension of the excavation process.	**Yes**: it preserves the details of the original hand drawing, but this operation is slow and has proven to be very problematic for data management and implementation.	**No**: It is not possible to use this method for adding additional information. The level of detail is set by the original hand drawing.	**Yes:** GIS has advanced editing tools for creating points, polygons or polylines to generate drawings using raster images as geometrical references.
Digitising orthoimages	**Yes**: it is possible to create and use orthoimages in the field (in a portable device) for digitising the contexts encountered during the excavation.	**Yes:** it is possible to improve the polylines and polygons. The graphic documentation can be enriched by adding new elements to the drawing.	**Partially**: this method allows digitising all visible elements in the orthoimage generating graphic descriptions of everything visible in the bidimensional projection.	**Yes:** GIS has advanced editing tools for creating polygons or polylines to generate drawings using raster images as geometrical references.
Total Station or RTK GPS	**Yes**: Total Station or RTK GPS can be used in the field to outline the contexts encountered during the excavation. In more advanced models it is possible to visualise the ongoing drawing on a screen.	**No**: these instruments can't be used in post excavation to change or increase the quality of the graphic representation.	**Partially**: the impossibility (with most models) to verify the quality of the graphic representations in the field makes this operation very difficult to perform.	**No**: the use of a pole for outlining the contexts reduces the possibility of maintaining control over the graphic representation.
Digitising 3D models	**Yes**: it is possible to create and use 3D surface models in the field (in a portable device) for digitising the contexts encountered during the excavation.	**Yes:** it is possible to improve the 3D polylines and polygons. The graphic documentation can be enriched by adding new elements to the drawing.	**Yes**: this method permits digitising all elements visible in the 3D model. The quality of details is directly related to the geometrical resolution of the model.	**Yes**: 3D GIS has advanced editing tools for creating 3D boundary models, 3D polylines and 3D polygons using 3D surface textured models as geometrical references.

In Kämpinge, 3D polylines were used for producing graphic documentation and recording the textual information at the trowel's edge. A geodatabase was designed to support single context recording and for including the 3D datasets created during the investigation (Figure 4.4). The possibility of producing graphic documentation on-site allowed us to perform more accurate observations and produce qualitative data to use during the investigation (see BOX 4.5). Since our first use of tablet PCs in the field, this technology now provides a much more advanced drawing experience, and if combined with the latest versions of 3D GIS platforms (e.g. ArcGIS PRO), can be employed to achieve accurate and comprehensive graphic descriptions of any geometrical feature visible on the model.

Although 3D boundary (solid) models are mostly generated during post-excavation activities, it is possible (if necessary) to interpolate 3D point clouds processed at the trowel's edge so as to create quick visualisations of the spatial relations between 3D representations of archaeological deposits. Current 3D GIS platforms permit importing high-resolution boundary models into the geodatabase in order to visualise them in spatial relation with the rest of the documentation recorded during the investigation (Figure 4.7). The generation of such data involves the use of additional software (see BOX 4.6) and the 3D representations resulting from this approach are derived from the interpolation of multiple datasets of information (e.g. 3D point clouds or 3D meshes), often recorder across different stages of the field investigation.

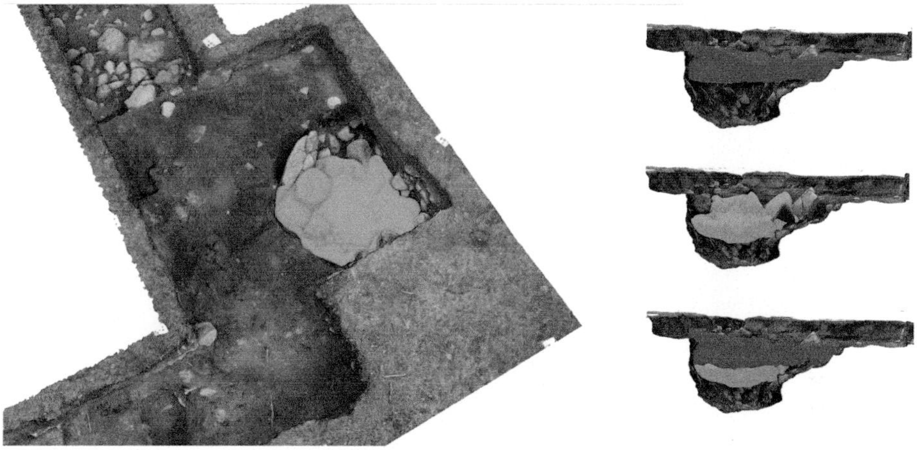

Figure 4.7 3D GIS platform developed as part of the Västra Vång archaeological field project. The image displays the implementation of a 3D boundary model derived from 3D surfaces in spatial relation with the rest of the archaeological documentation. Source: Paola Derudas, image courtesy of the Västra Vång Project.

BOX 4.6 3D boundary representation of archaeological deposits

The methodology described in this box is developed using Agisoft Metashape, CloudCompare and ESRI ArcGIS Desktop/PRO. The 3D surface models imported into the GIS were acquired using a DSLR digital camera; the images were generated using Metashape 1.7. and further processed using CloudCompare for creating the 3D boundary models. ArcGIS PRO was used for the visualisation.

This BOX will provide a general overview of how to create and import into the 3D GIS boundary models derived from the interpolation of multiple 3D point

(Continued)

clouds. When using image-based 3D Modelling or laser scanning for recording the sequence of contexts encountered during the field investigation, it is possible to interpolate the 3D surface models that represent a sequence of two contexts and thus generate a 3D boundary representation of the archaeological deposit removed during the excavation.

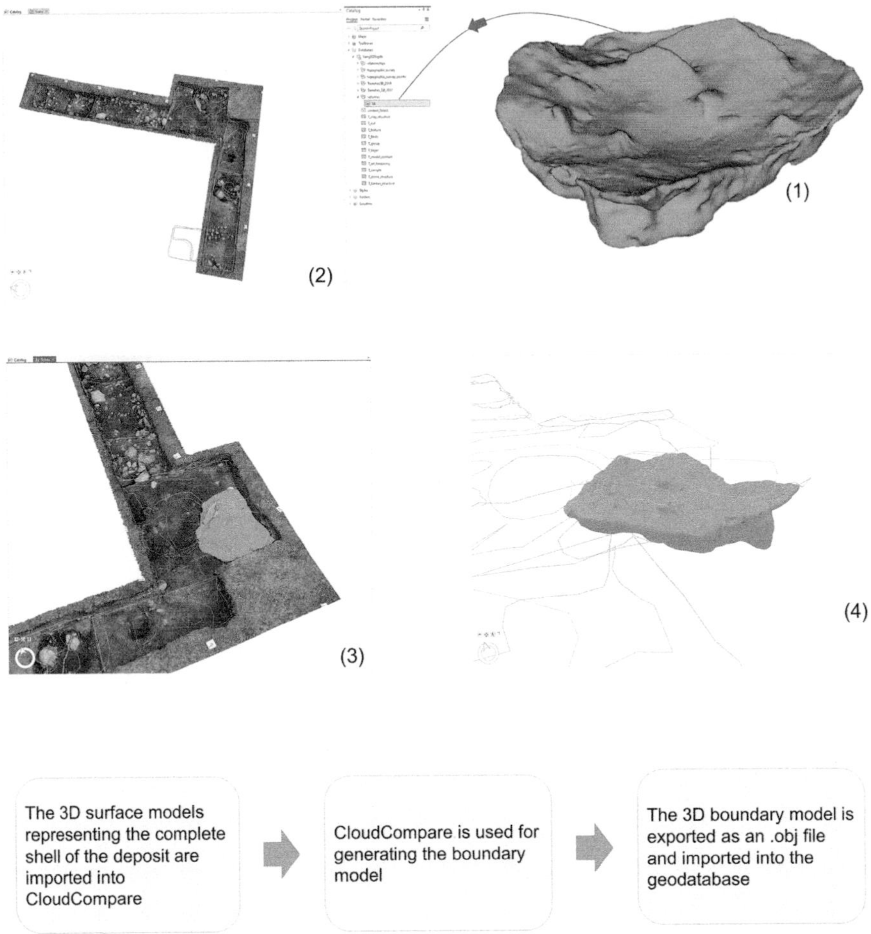

The 3D surfaces (or 3D point clouds) generated during the field investigation (created with Agisoft Metashape, in this case) are imported in a 3D point cloud processing software (e.g. CloudCompare), to be merged and re-processed in order to obtain a boundary representation of the deposit. Once in CloudCompare the point clouds (constituting the model surfaces) can be optimised (cloud subsampling), merged and, once the normals have been re-computed, a Poisson surface reconstruction is used to generate a new surface (**1**) (see Roosevelt et al., 2015). The 3D boundary model is exported as an obj file and imported into the 3D GIS geodatabase (**2**) as a multipatch file (see BOX 4.4) (**3**). The 3D object is stored together with the rest of the 3D documentation and can be visualised in spatial relation with other 3D boundary models (representing different deposits) (**3,4**) (Figure 4.7). 3D models characterised by a topologically consistent surface can be used to extract accurate information about the volume and can be linked to an attribute table incorporating information coming from different specialists operating on-site (e.g. archaeobotanists, geologists, paleo-botanic data; pottery; osteological remains).

Once in the geodatabase, 3D boundary representations of the deposits can be easily grouped into a multipatch feature class and used to connect each unit of volume information with database entries. This is particularly useful for generating (through query operations) 3D distribution or density maps related to different categories of data collected by the specialists involved in a field excavation.

A few notes: 3D boundary representations of archaeological deposits are 3D records derived from the interpolation of 3D point clouds acquired during distinct phases of the excavation. These types of 3D representations are often visualised in the 3D GIS using false colour, and the successful production and implementation of such models strongly depends on the quality of the field documentation carried out in the field.

Boundaries are not always easy to generate (much depends on the geometrical complexity of the deposits encountered in the field), and the 3D point clouds used for the generation of boundary representations are often subject to intense post-processing work. However, they allow archaeologists to (a) review the topological relations between deposits, and thereby gain a clearer understanding of the stratigraphic sequence encountered during the investigation, and (b) if converted into a multipatch file, imported into a geodatabase and linked to an attribute table these records can incorporate and display the work of different specialists, providing the opportunity to perform advanced spatial analysis. Additional and complementary methods to obtain and describe volumetric information related to archaeological stratigraphy are further described in Chapter 7.

4.2.3 Shifting grounds

The methodology defined during the field investigation at Kämpinge demonstrated how 3D GIS in the field can go far beyond simple visualisation/interaction by promoting a heuristic approach to the data encountered. The possibility of simulating and discussing multiple scenarios at the moment of engagement with material remains from the past demonstrates the capacity of these platforms to challenge theories and ideas at the level of tangible materials (Edgeworth 2014) (see discussion Chapter 2).

4.3 Integrating the work of specialists operating on-site

3D GIS can be used to identify and address new research questions. This subchapter focuses on the experience developed within the framework of Sandby borg; a ring fort located on the island of Öland Sweden, in which a 3D GIS was used to visualise the combined activities of the field archaeologists and osteologists operating on-site; the use of 3D GIS allowed performing a new type of analysis gaining information which were impossible to detect using a more traditional approach.

4.3.1 The Case of Sandby borg

The island of Öland off the east coast of Sweden hosts several ringforts dating to the Scandinavian Iron Age. These structures have been investigated over the years in several field projects. One of the most recent investigations took place at Sandby borg, a ring fort located 2 kilometres southeast of the village of Södra Sandby, Sweden. In 2011 a field investigation revealed signs of a dramatic event which took place during the Migration Period (c. AD 400–550) which caused the abandonment of the site. The ringfort is dated to the late fifth century and was excavated by the Department of Museum Archaeology from Kalmar County Museum. Over the investigation seasons, the remains of several people, slain and left where they had fallen, were retrieved. The archaeologists began to investigate the relationship between these human remains and the structures, and thereby gained a unique perspective on domestic life during the Migration Period (Alfsdotter et al. 2018:421).

During 2012 and 2013, the Department of Archaeology and Ancient History at Lund University developed an experiment in collaboration with the Kalmar County Museum which aimed to establish an investigation method capable of combining spatial analysis and osteological research (Wilhelmson and Dell'Unto 2015). The human remains were identified in 2011 and excavated during the field campaigns carried out in 2012 and 2013. During both investigations, image-based 3D Modelling techniques and RTK GPS were used for creating accurate and high-resolution 3D spatial records of the contexts and features retrieved during the study. 3D GIS was used for visualising the archaeological records detected during the field campaigns using a data implementation pipeline which employed the commercial package Agisoft Photoscan 1.0 for the generation of the 3D models, Meshlab for the inspections of the models in high resolution and ESRI ArcScene 10.1 for the spatial analysis. When this experiment was

Figure 4.8 3D simulation of house 40 containing individual 1 (on the right side of the image) excavated during the 2012 season, individual 2 (on the left side of the image) excavated during the 2013 season and the partially excavated structures, along with the artefacts retrieved during the 2013 season (on the upper part of the image) (Wilhelmson and Dell'Unto 2015).

developed, the visualisation and editing tools offered by these platforms were limited. Today, a similar approach would be easier to implement and could lead to clearer and more consistent results. Despite these limitations, this approach allowed us to gain a clear and accurate visualisation of the relations between the different contexts and features identified during the investigation. It provided an accurate understanding of the spatial distribution of the finds in relation to the events connected with the destruction and abandonment of the site (Figure 4.8).

The approach adopted for developing this project is not dissimilar to the ones described within the previous sections, but it is unique in the opportunity it created for tight collaboration between field archaeologists and osteologists. It gave us a chance to customise the 3D GIS for supporting advanced spatial and osteological analysis and identify archaeological evidence that was only visible in the 3D space.

After a closer inspection of the human remains in the laboratory, the information retrieved was imported directly into the 3D GIS and connected to 3D polylines and polygons used to identify – directly on the 3D models – their correct spatial location in the virtual scene. In order to include these results within the 3D GIS the geodatabase was customised for hosting the information retrieved in the laboratory.

The use of 3D GIS allowed us to gain a clear and accurate overview of the spatial relations among the different remains in relation to the different phases of the excavation, and during the analysis carried out in the laboratory (Wilhelmson and Dell'Unto 2015).

The possibility of contextualising the work carried out in the lab and comparing it with the different finds detected during the investigation in such great detail represented an incredible opportunity for contextualising the post-excavation work within a diachronic visualisation of the excavation sequences. The results of the spatial analysis highlighted a clear connection between the archaeological structures and the fractures identified on the bodies, allowing the archaeologists to identify patterns that would have been impossible to detect in any other way (Figure 4.9).

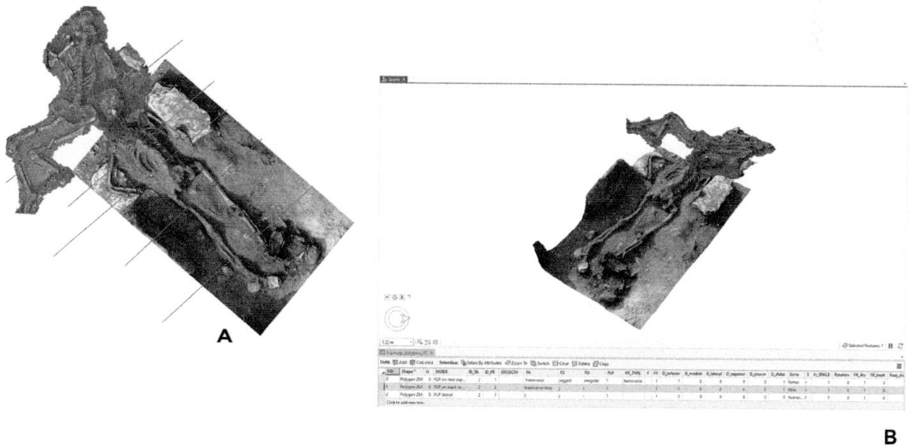

Figure 4.9 The 3D models of the individuals identified during the field campaigns in 2012 and 2013, visualised in the 3D GIS after the analysis carried out in the laboratory. Image (A) reveals the pattern, identified in the GIS using a specific set of queries highlighted using 3D polylines, in a group of fractures that occurred in relation to a single event, probably the collapse of the house roof. Image (B) displays the customised geodatabase used for hosting the information identified by the osteologist in the laboratory (Osteological Analysis performed by Helene Wilhelmson) (Wilhelmson and Dell'Unto 2015).

BOX 4.7 Implementing 3D surface models of human remains in the 3D GIS

The methodology described in this box is developed using ESRI ArcGIS Desktop/PRO; the 3D models imported in the GIS were acquired using a DSLR digital camera; the images were processed using Metashape 1.7. ArcGIS PRO was used for the visualisation and analysis.

This BOX describes the steps needed for visualising and working with 3D models (derived from the use of image-based 3D Modelling techniques) in order to include human and animal remains within a 3D GIS platform. To import the 3D model, it is sufficient to follow the steps described in BOX 4.4, but due to the geometrical complexity that characterises this type of remains, it is essential to thoroughly clean the scene (physically) and acquire pictures from additional angles. This will ensure that the software has enough information to perform a comprehensive description of the geometrical characteristics and disposition of the bones in the scene. Today, several image-based 3D Modelling products can provide an immediate assessment of the quality of the points projected in the scene (e.g. Metashape/Dense Cloud confidence); these tools are very useful and permit increasing the quality of the model by adding (when possible) additional pictures to the dataset. Recording osteological data often requires a higher number of pictures for the representation of a single scene. The use of smaller markers is often adopted (a) for keeping a higher control on the camera optimisation, (b) for facilitating the georeferencing process or (c) for recording scattered bones (**1**). The information identified in the laboratory after a closer inspection of the bones can be recorded in the 3D GIS using 3D polygons, polylines or points (Figure 4.9).

In order to visualise pieces of evidence identified during different phases of the excavation in the same virtual scene, it could be necessary to select from the 3D models all the polygons which do not directly refer to the osteological material and remove them. This operation can be performed directly in the image-based 3D Modelling software before exporting the models (**2**).

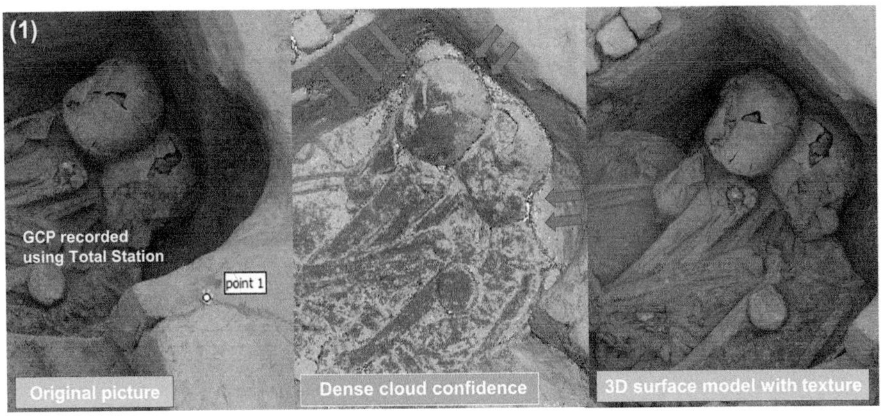

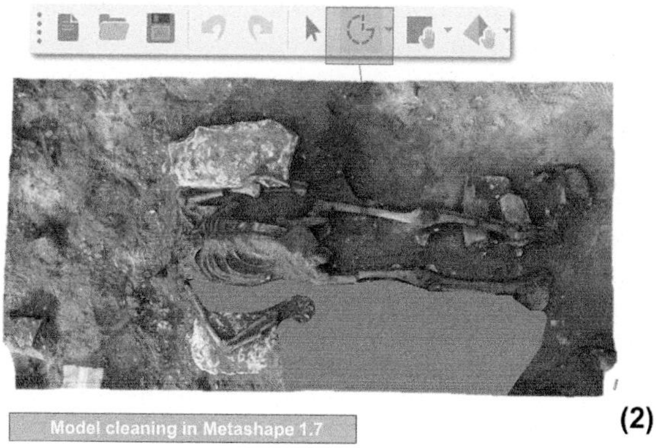

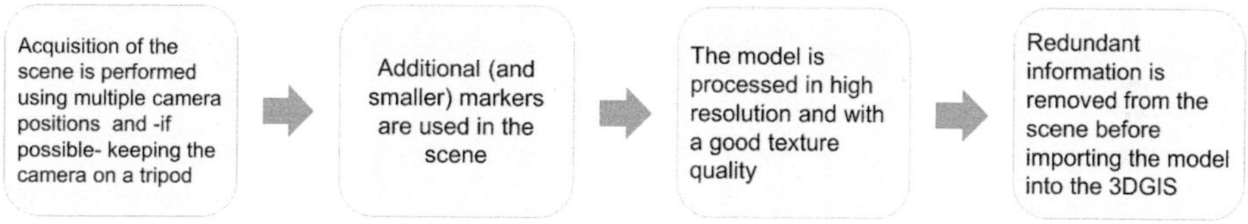

A few notes: It is important to keep in mind that using image-based 3D Modelling or laser scanning will provide 3D models representing only the visible surface of the bones on the ground, and that for obtaining a complete and clear view of the distribution of the human or animal remains across the scene, it might be necessary to merge the different 3D objects recorded during different steps of the investigation in a single layer. This could be achieved in two different ways: either by (**A**) aligning all the bones identified during the field campaign one by one using an external software (e.g. Metashape or CloudCompare) and then importing groups of bones into the geodatabase as a "single" entity, or (**B**) by importing single georeferenced models (each representing a single bone) into the 3D GIS, to be virtually reassembled later in the 3D scene (e.g. merging them all in one multipatch file). Both strategies have their particular strengths and weaknesses, and their adoption depends on the visualisation goals of the project. Another important aspect to keep in mind when documenting human and animal remains relates to the type of techniques adopted for recording the GCPs. It is recommended to use the total station's prism directly on the GCPs (removing the pole). This will reduce the possibility of generating errors caused by accidentally moving the pole.

4.5 Summary and conclusions

This chapter focused on the use of 3D GIS for supporting field activities carried out during various investigations. By using examples from different projects, the text provides several considerations concerning the impact that 3D GIS is having on archaeological field practice. The chapter contains several informative BOXes designed to provide general instructions on how to implement and use 3D models derived from 3D recording techniques in the 3D GIS environment. The last part of the text provides an example of how 3D GIS can be used to include the work of specialists operating in the field, identifying new research methodologies and defining new directions.

This chapter underlines the impact that archaeological practice has on the process of data interpretation, and reflects on how its transformation requires a long experimentation period in which archaeologists engaging with new technologies test new approaches for defining new field practices. Furthermore, Chapter 4 stresses the importance of establishing data acquisition routines that allow archaeologists to assess the records acquired during the investigation against the physical evidence encountered in the field. Using the 3D GIS for simulating and discussing different interpretations in the field demonstrates the capacity of these platforms to challenge theories.

The next chapter will discuss and present examples of how to use 3D GIS for supporting surface and subsurface analyses.

5 Surface and subsurface analysis

5.1 Introduction

An interesting area of application for analytic 3D GIS is represented by surface and subsurface assessment of threatened archaeological/cultural heritage sites. The recent advances in the acquisition techniques (Callieri et al. 2011; Dellepiane et al. 2013; De Reu et al. 2014; Forte et al. 2015) now make it possible to take advantage of multiple methods for documenting the state of preservation of "at risk" features. The dramatic increase of image-based 3D Modelling techniques has led to a massive use of 3D models for archaeological site documentation. As we have already mentioned, surface 3D models represent an important resource for reproducing the macro characteristics of virtually any type of archaeological context with a high level of detail: positive and negative units (see BOX 4.1), horizontal and vertical surfaces, anthropogenic or natural layers. In this sense, changes in surface morphologies can be documented and mapped by relying on a multi-temporal approach to data acquisition that allows specialists to monitor the state of preservation of an archaeological feature on a regular basis. Collapsing buildings, cracked walls, or soil assemblages are only a few examples of heritage at risk that requires innovative strategies for the damage assessment. Notably, factors of risk are manifold and can mainly be characterised as being of human-made or environmental origin. Although traditional GIS-based approaches still constitute a crucial option for mapping and assessing the state of preservation of heritage sites on a large scale, it is important to consider additional options when evaluating the damages to single monuments or archaeological features. This involves the use of georeferenced 3D boundary/surface models and a thorough examination of any small-scale threatening factor.

In this chapter, the case studies show different modes of approach in which a combination of 2.5D/3D analytic tools were employed for assessing the state of preservation of different kinds of archaeological material.

5.2 Evaluating the damage in a Viking Age site

This case study concerns an archaeological site located in the small island of Öppenskär, in the Blekinge region of south-eastern Sweden (Figure 5.1).

Despite limited knowledge about Öppenskär island, we have good information about the eastern archipelago of Blekinge and its strategic importance as a major trading centre during the later part of the prehistoric period.

The site itself lies on one of the most topographically prominent points in the archipelago with a very high degree of visibility over the surrounding area. Its discovery took place during illegal metal detecting activity reported to the local museum of Blekinge. After a preliminary assessment of the site chronology, a more thorough examination

DOI: 10.4324/9781003034131-6

took place as a consequence of destructive action caused by a vehicle used for woodland clearance, resulting in damage affecting the topmost layers of the site stratigraphy.

In this case study, 3D GIS was tested as a method of damage assessment at this Viking Age site as a result of human-made destructive action. The examined area, interpreted as a Viking Age grave and/or silver hoard site, was investigated after the partial destruction caused by the vehicle rampage. To make a thorough assessment of the damage, a field campaign was conducted in order to assess the level of surface erosion affecting the site in quantitative terms, but also to gain a deeper insight into the chronology and the function of the site.

Image-based 3D Modelling was employed to generate surface models that formed the basis of a quantitative evaluation of the erosion caused by the wheels of the vehicle on the archaeological site. Such a pipeline can be potentially extended to different research contexts and is mainly intended as an additional instrument in support to site preservation and rescue archaeology. Indeed, the novelty of such an approach is that it can provide

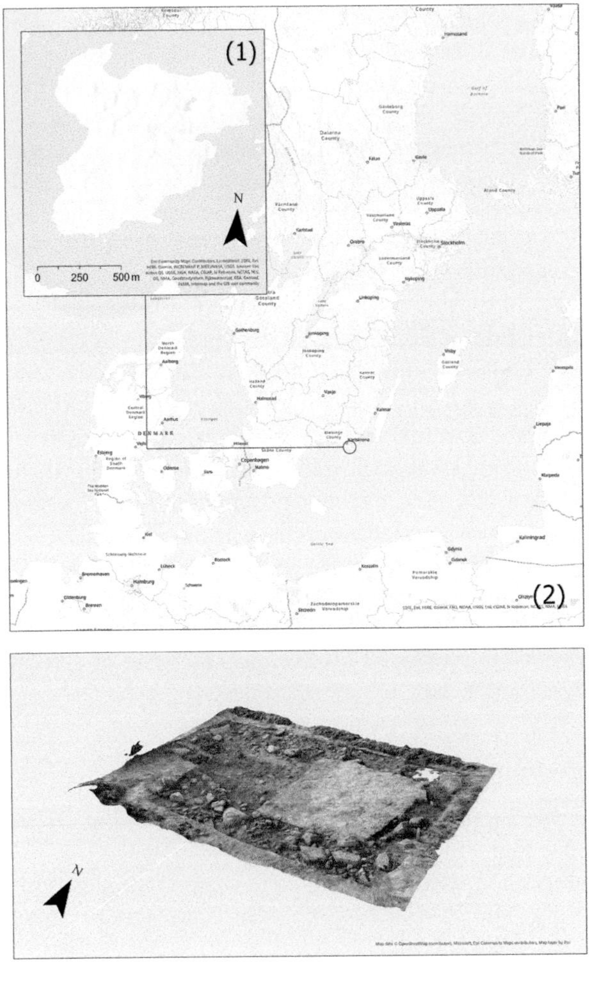

Figure 5.1 Öppenaskar island, located in the Eastern Archipelago of Blekinge, South-Eastern Sweden. Below, a 3D surface model of the described excavation area (Landeschi et al. 2016).

local authorities and specialists with a quantitative assessment of the amount of archaeological material that has been damaged by a destructive anthropogenic action.

Just before the excavation took place, a series of photographs were acquired by archaeologists in order to document the condition of the ground surface after the passage of a vehicle in an area that was previously interpreted as belonging to a Viking Age site (Henriksson and Nilsson 2015). All these photographs were used to produce a 3D surface model of the area under investigation and the idea was to compare it with 3D models produced at a later stage after the stratigraphic excavation was started. In this way, it would be possible to georeference the surface models covering the same portion of terrain and assess the degree of terrain erosion caused by the destructive vehicle action and quantitatively estimate the damage suffered by the archaeological features.

5.2.1 Field data collection

Different phases of the stratigraphic excavation were documented in situ, with the acquisition performed through a combined use of DSLR cameras, laptop and tablet computers (Landeschi et al. 2016). By taking advantage of dedicated software for image-based 3D modelling, it was possible to create separate project files and define different levels of spatial resolution to fulfil different purposes for each surface model. 3D models of the trench area were georeferenced with an RTK GPS and imported into 3D GIS following a pipeline similar to the ones described in previous chapters.

Managing the data in a georeferenced space was particularly useful due to the analytic needs of this project. As previously mentioned, one of the goals was to make an assessment of the damage that occurred to the surface of the site. For this reason, it was important to place the single 3D surface model of the ground surface documenting the destructive vehicle action in the correct spatial relation with the 3D model documenting the first archaeological layers uncovered after topsoil removal.

This approach, which is thoroughly described in BOX 5.1, can be extended and applied to many other contexts. Nowadays, 3D surface models can be acquired relatively easily and can be used to document surface preservation conditions on-site on a regular basis, and thus monitor all the changes caused either by anthropic or natural factors. 3D surface models can provide us with different solutions for extracting and analysing geometrical information. Vectors and rasters, Digital Surface Models (DSMs), and 3D meshes represent data formats that can be adopted to perform map algebra calculations or vector-based Boolean tasks where topological relations between different models are examined.

BOX 5.1 Combining raster-vector data analysis to assess surface erosion

The methodology described in this box is developed using ESRI ArcGIS Desktop/PRO version; the 3D surface models imported into the GIS were acquired using a DSLR camera and processed using image-based 3D modelling techniques, from Agisoft Photoscan/Metashape software.

To assess the issue of surface erosion often associated with archaeological damages and architectural degradation, it is possible to combine vector and raster datasets and to perform analytic tasks following a well-defined workflow (**1**).

(Continued)

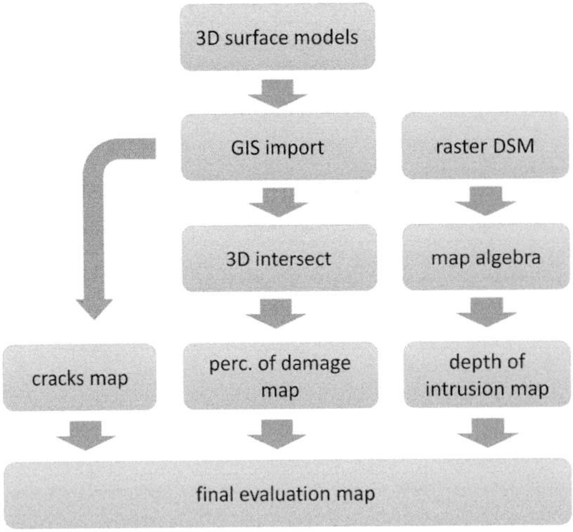

In this case study three different 3D surface models were imported as multipatch feature classes into GIS. Regarding the 3D model setup, each was processed in Agisoft Photoscan/Metashape in order to provide a sufficiently accurate representation of the surface geometry (approximately 500K vertices).

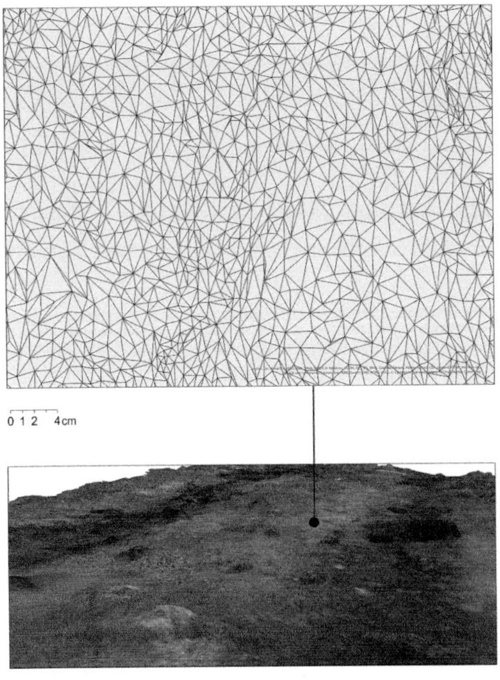

In particular, spatial resolution of the meshed models was set to relatively high levels of point density (1pt/cm) to improve the accuracy of the assessment and better define those portions of spatial overlap resulting from an erosive process (**2**).

As a first surface 3D model (TR00), a multipatch feature class representing the ground surface gouged by the vehicle's wheels provided us with elevation values of the terrain. One more 3D surface model (TR01) was produced to represent the same portion of terrain after the stratigraphic excavation had begun and the topsoil had been removed. Once all the 3D models were correctly georeferenced, it was possible to perform analytical tasks involving topological and algebraic operations between the examined surface models. The key issue was calculating any existing intersection between TR00 and TR01, assuming that any "eroded" portion of surface affecting the archaeological deposits beneath and recorded in the uppermost model, would have resulted in an elevation value lower than the corresponding portion of surface in the lower model (3).

Areas of the ground surface model with a lower Z value than the corresponding area in the lower model (archaeological layers) would very likely correspond to the portion of the archaeological site eroded by the wheels (modified after Landeschi et al. 2016).

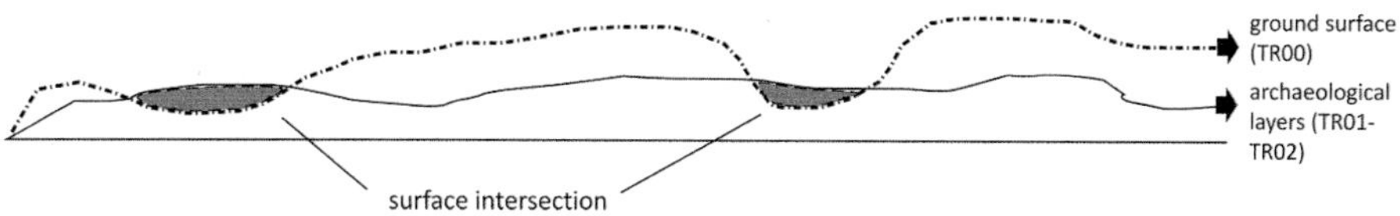

By using the 3D Intersect line function, a 3D shapefile marking the intersection between the overlapping surface models was obtained. The sectors enclosed within the intersecting lines corresponded to those portions of terrain affected by the destructive action of the wheels. To restrict the analysis area to the portion of terrain affected by the wheels' action, a further vector layer was added. This layer was converted into a 3D polygon shapefile and then intersected with the shapefile resulting from the previous intersection between TR00 and TR01. This operation allowed us to define a contact surface between the two surface models so as to define the extension of the area affected by the wheel action.

This part of the workflow can be summarised as follows: (**4**) 3D Surface models representing the ground surface after the destructive action of the wheels (TR00, to the left) and the removal of the topsoil during the stratigraphic excavation (TR01, to the right); (**5**) intersection between the models allowing us to define the portion of terrain affected by the wheel action as a 3D line shapefile (**6**) (modified after Landeschi et al. 2016).

A 3D polygon shapefile was then derived, characterised by two categories of polygons and expressing intersection or no intersection, respectively, between the examined surface models (**7**)

(Continued)

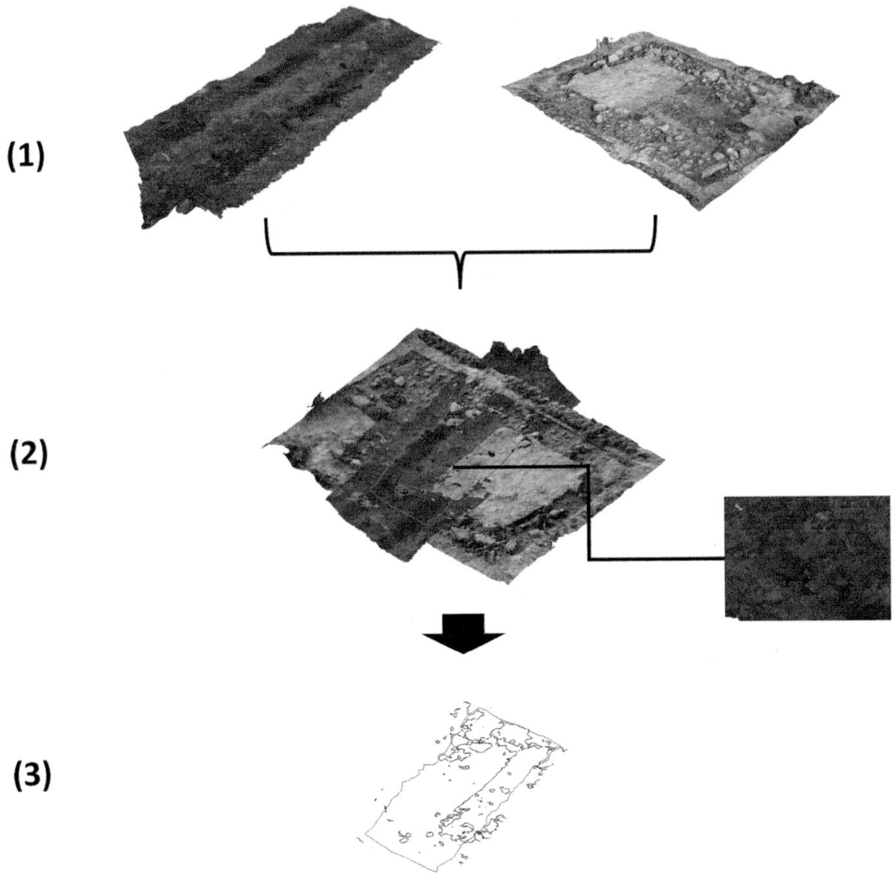

In particular, the 3D line shapefile was converted into a 3D polygon shapefile and used to calculate the status of the terrain under investigation in a binary way. Red values represent areas of intersection between models TR00 and TR01 and, therefore, are probably affected by the wheel action. This stage of evaluation allowed us to determine the percentage of the area crossed by the vehicle in which archaeological material was damaged (34%).

The binary evaluation between affected and non-affected areas had to be further refined by introducing a more accurate evaluation of the depth of penetration exerted by the vehicle wheels into the ground. To do so, 3D surface models were converted into 2.5D raster DSMs. Such an operation allowed us to assign an elevation value to each single spatial unit (pixel) belonging to the terrain models. Map algebra operations were performed to calculate the algebraic difference between overlapping pixels belonging to TR00 and TR01. The resulting raster map expressed a numeric value corresponding to the difference in elevation between point "n" in TR00 and point "n1" in TR01. Negative numbers indicated a contact/intrusion of the upper layer into the one beneath, indicating, in this case, a gouging action affecting the archaeological deposits.

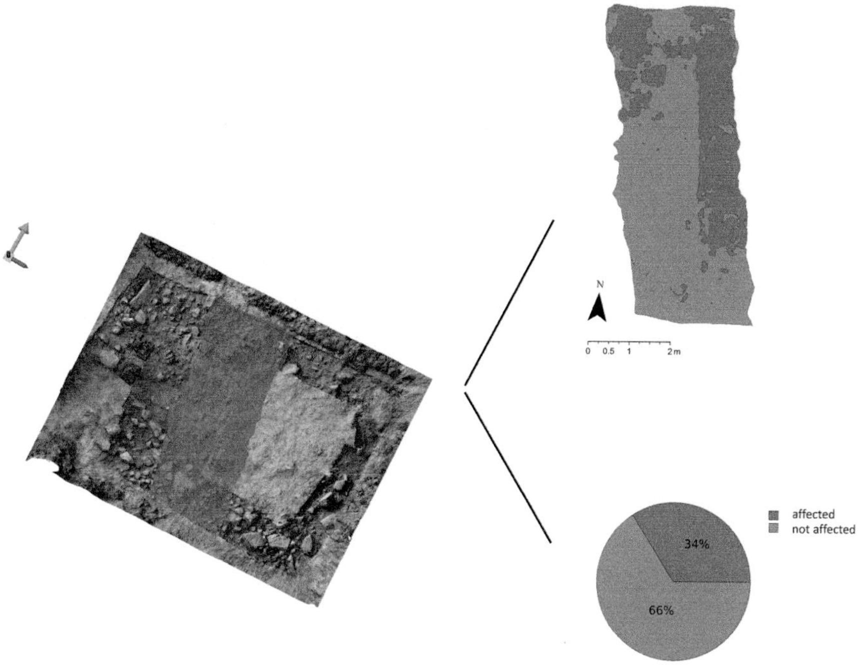

To summarise this part of the workflow (**8**): (A) DSMs were created to represent elevation values for TR00 and TR01; (B) by applying map algebra functions it was possible to calculate the difference in elevation between the two examined surface models on a pixel basis. By overlapping the obtained map and the binary assessment (C), it was possible to determine the depth of intrusion within that portion of terrain previously marked simply as "affected" (D).

As a final outcome, depth of surface intrusion exerted by the uppermost surface model into the model of the archaeological layers was calculated. A range of information was associated with this map: the amount of surface exposed to the destructive action of the wheels, expressed in percentage values; a calculation of the quantity of surface units eroded by the wheels, based on the depth of surface intrusion; and a final map allowing us to put the distribution of observed stone cracks in spatial relation with the areas previously marked by higher levels of depth intrusion (**9**). Notably, there we observed a good match between the two features.

(*Continued*)

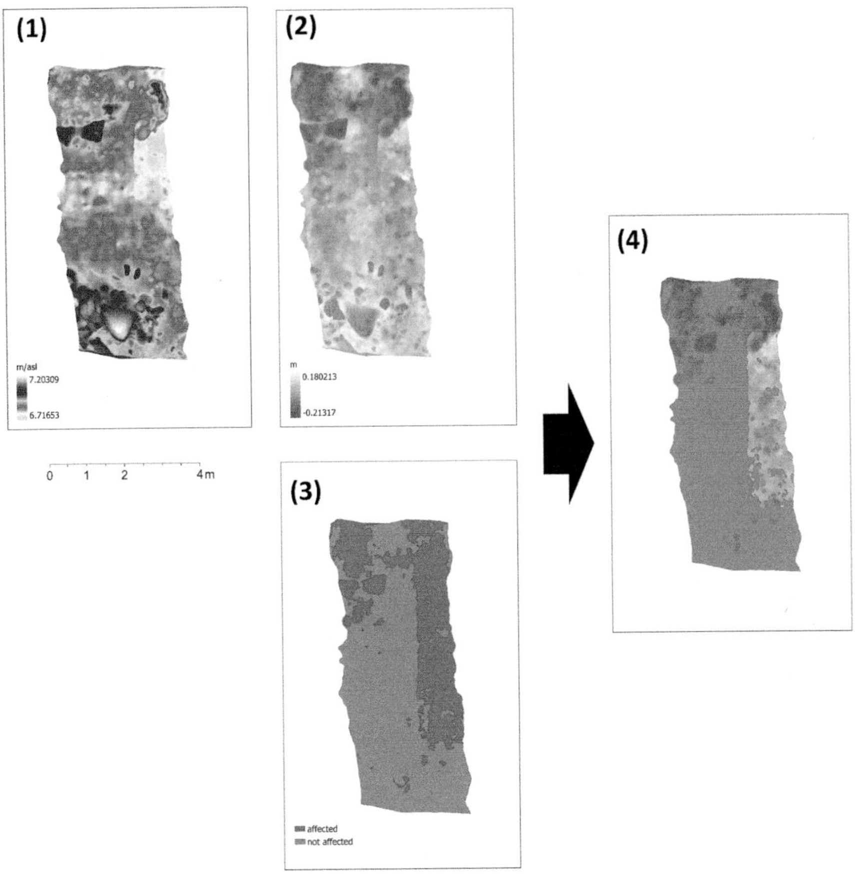
(1)
m/asl
7.20309
6.71653
(2)
m
0.180213
-0.21317
(3)
affected
not affected
(4)
0 1 2 4m

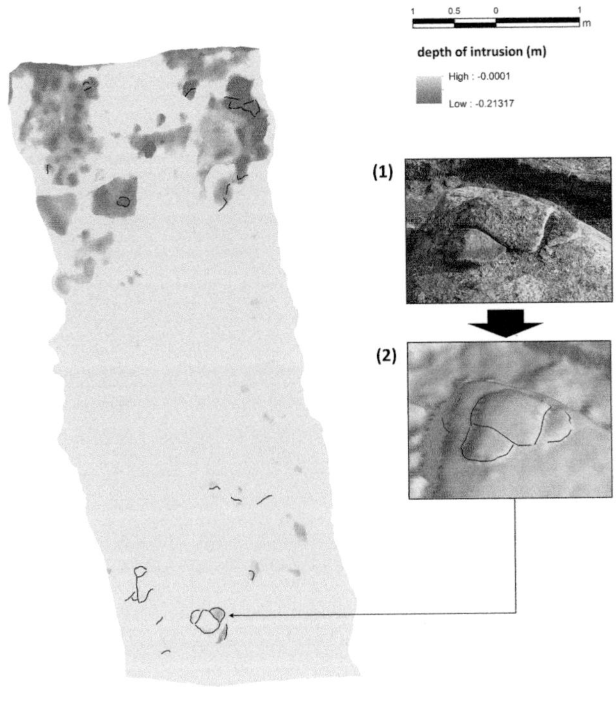
1 0.5 0 1 m
depth of intrusion (m)
High : -0.0001
Low : -0.21317
(1)
(2)

> **A few notes:** it is important to generate a dense point cloud with a sufficient point density for creating a meshed model with a good spatial resolution. The smaller the expected size of the crack patterns (or any other damage trace), the higher the resolution of the model should be.

5.2.2 Results and discussion

Drawing upon the obtained results, an assessment of the level of damage to an archaeological stratigraphy can be obtained by combining spatially high-resolution 3D surface models, derived from image-based 3D Modelling, with analytical tools available in GIS. Due to the complexity that often characterises stratigraphic relationships between contexts, it is crucial to rely on more advanced techniques of data representation which include the so-called "Z factor". Although the proposed workflow does not use a strictly fully 3D spatial analytical approach, the use of Z values in the form of attributes of generated DSMs provided the necessary opportunities to perform calculations that took into account any minor variation in the elevation of the examined surfaces. Such variation is an important indicator of changes caused by destructive actions of intrusion, which in the case of Öppenskär island could be linked to the vehicle rampage. Similar workflows to the one described here can be extended and applied to support land evaluation practices and cultural resource management. The damage assessment aspect is particularly relevant nowadays all-around Europe due to the increasingly dominant role of rescue archaeology as the most common form of field practice. Adopting formal methods of assessment could provide specialists with additional means for documenting and reporting the state of preservation of heritage in areas affected by environmental and human-made threats (Landeschi et al. 2016).

5.3 Pompeii risk analysis

The second case study explores the use of a 3D GIS for supporting the work of conservators. This experiment was developed as part of the research activities of the Swedish Pompeii Project (SPP), in which the system was used for the identification of structures at high risk of collapse. The results of this work were published in Campanaro et al. 2015. The experiment was applied to the structures of the North House of Caecilius Iucundus and the methodology was customised (a) for gaining accurate assessments of the decay process of the structures, (b) for characterising the 3D models of the structures with information such as crack patterns and masonry singularities, (c) for generating risk maps of the standing walls, and (d) for identifying the right actions to undertake for supporting the conservation process (Figure 5.2) (Campanaro et al. 2015).

By implementing 3D thematic maps generated as the results of multiple analyses in the geodatabase management system (GDBMS), it was possible to customise the system so that it allowed conservators to pose specific queries about the overall status of the structures. The creation of these thematic maps is a process based on the identification, selection, and extraction of specific elements – usually described by the conservators as single layers or entities – that can be used (or combined) to assess specific risk situations connected with the status of walls and structures.

The 3D models of the single walls were processed in the GIS to extract specific information (e.g. loss of verticality, degradation, etc.) which was later linked to the models

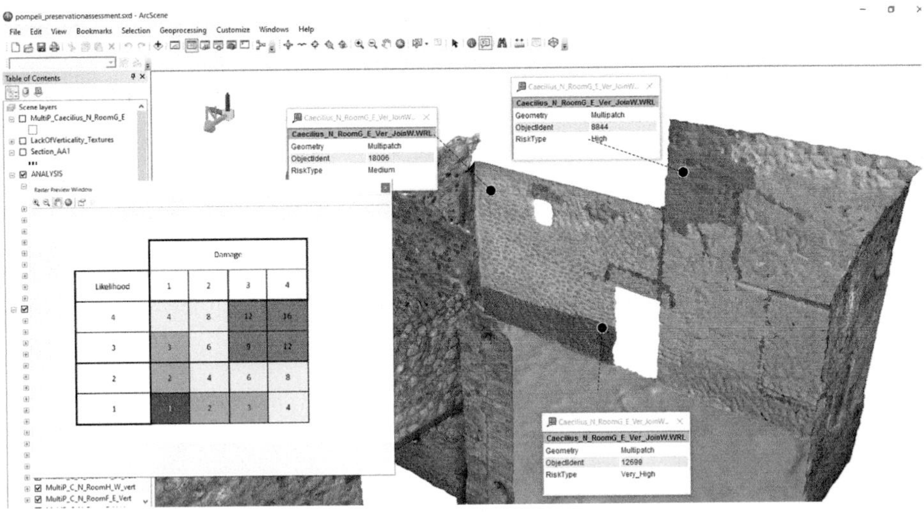

Figure 5.2 Example of a risk map generated using the 3D GIS. Courtesy of Danilo Marco Campanaro (Campanaro et al. 2015).

and included in the geodatabase. Unfortunately, due to the current 3D GIS platforms' limitations regarding spatial analysis on vertical surfaces, the generation of thematic maps was possible only through the development of a pipeline which involved the rotation of the vertical structures onto a horizontal plane. This allowed us to use 3D analysis tools (originally designed for analysing horizontal surfaces) to highlight different aspects of the walls' geometry. These results were automatically transferred by the software to the attribute tables of the original models and used for assessing the wall's status from multiple perspectives.

This information allowed the team to create decay (vector) and risk (raster) maps and highlight the loss of verticality or any other specific (usual or unusual) geometric element visible on the walls. The use of thematic maps and the extension of the attribute tables meant that the 3D GIS could be used for the generation of different risk levels, estimated by combining specific types of deterioration in relation to the severity of their damage (Campanaro et al. 2015:330).

BOX 5.2 Creating and displaying thematic maps on the 3D model for supporting the work of conservators in the 3D GIS

The methodology described in this box is developed using ESRI ArcGIS Desktop/PRO version; the 3D models imported in the GIS were acquired using laser scanning; the point clouds were post-processed using Meshlab for generating the mesh and Blender for projecting the textures.

As previously described, current 3D GIS platforms do not always provide analytic tools capable of dealing with vertical surfaces, and for this reason, in order to create thematic maps for highlighting specific characteristics of the structures directly in the GIS it is necessary to set up a pipeline which uses the 3D analysis tools designed for operating on horizontal surfaces.

This is achieved by cloning and rotating the vertical surface onto a horizontal plane (**1–2**). This can be performed directly in the GIS using the editing tools available for rotating geometries or by exporting the 3D models into a software dedicated to 3D visualisation or modelling such as Blender, computer-aided design (CAD) software applications like AutoCAD or 3D point cloud and mesh processing software like Cloud Compare.

Once rotated, a raster map is generated and used to describe different aspects of the structures' surface, such as leaning walls, or to identify fractures on the walls (**2**). The colour information corresponding with the results of the different analysis (and associated with each triangle composing the surface) is automatically updated on the original vertical surface using the triangle's unique ID.

(1)

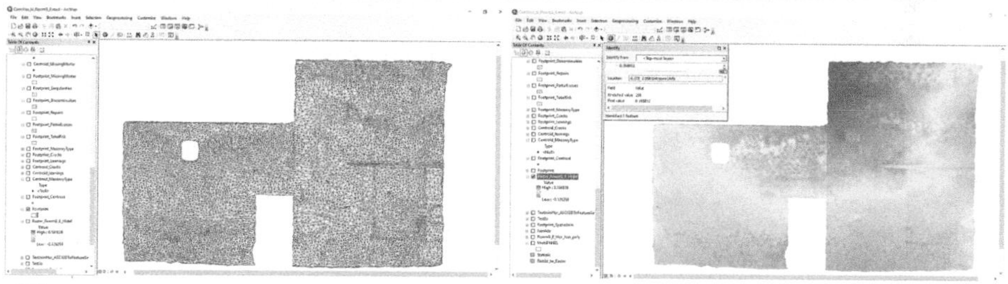

(2)

This process can transfer the information gained on the horizontal clone directly onto the attribute table of the original (vertical) model (**3**), providing the operator with the possibility of performing multiple queries for assessing the status of the walls from different perspectives (Figure 5.2). Or like in the specific example used in this BOX, the identification of the least vertical walls that are consequently at the highest risk of collapse (**3–4**).

(Continued)

This workflow is based on synchronising values from the horizontal raster surface to the corresponding vertical meshes (triangles). A footprint layer is created from the horizontal clone of the vertical multipatch. Subsequently, a centroid layer (feature to point tool) is generated for each triangle of the footprint. The horizontal clone is therefore used to develop 2D maps and receive the related attributes. Values from thematic maps are thus passed to the footprint via the centroids. This information is then transferred to the vertical clone based on the correspondence between the elements (ID) of the clones of the same facade. A major advantage of this method is the real-time updating of attributes. In fact, a change of a value in a feature sub-element in the 2D layers (footprint) would affect the corresponding 3D multipatch sub-elements (Campanaro et al. 2015).

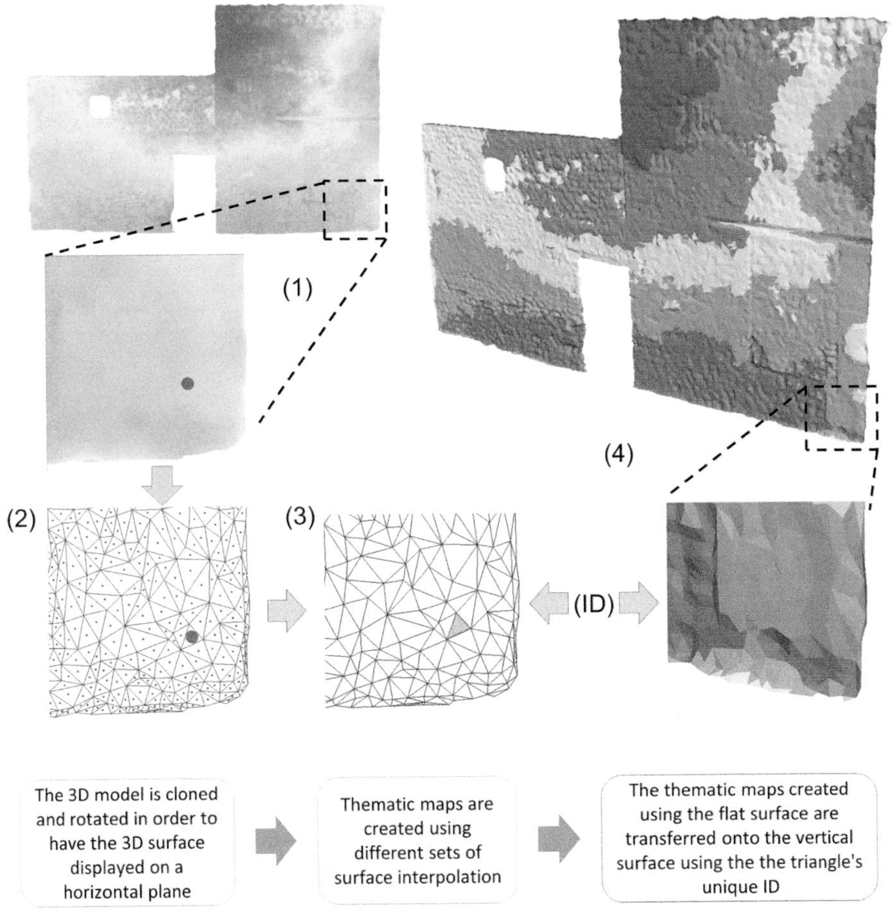

A few notes: a clear limitation of this method is the process of changing the orientation of the 3D surface; it would indeed be more practical to work directly on the model in its original position. Nevertheless, this approach proved to be an efficient workaround for identifying and combining tools that could track structure decay, create risk maps, or assess the verticality of standing structures. Once visualised in the vertical surface, these maps can be used as backdrops for defining additional layers identified as a result of combined queries.

The same pipeline can be easily applied for creating thematic maps for highlighting features on vertical surfaces such as rock art, runestones, and so on.

5.4 Summary and conclusions

This chapter provided an overview of some of the possibilities connected with the use of 3D GIS for the analysis of archaeological surfaces and subsurfaces. The case studies presented in the chapter were chosen to provide the reader with a broad and wide-ranging overview of the possibilities connected with the use of these methods in different scenarios. Differently from the previous chapters, where the BOXES are characterised by more straightforward pipelines, the methods described in this chapter imply a more advanced knowledge of the tools available in GIS and a deeper understanding of the functions offered by the system. These examples (as well as others described in this volume) are meant to stress how 3D GIS can be used for challenging theories and ideas during the engagement with tangible materials, rather than afterwards. They therefore highlight the role of these systems as bridging material and immaterial records. The next chapter will stress these aspects and suggest examples of how 3D GIS can be used to assess the visual impact of features, either in the landscape or in architectural space, by employing an entirely 3D method performed in a georeferenced space.

6 3D visibility analysis

6.1 3D-based visibility studies in archaeology: an overview

Visibility studies are a form of investigation in which human sight is simulated in order to evaluate what kind of impact a (pre)historic monument (or any other archaeological feature) would have had on a hypothetical set of observers. As has been noted in Chapter 1, visibility studies have a longstanding tradition in archaeology, with many examples of landscape-based studies involving the use and combination of GIS (Geographical Information System)-based tools for quantitative analysis of the viewshed associated with one or more anthropic elements (hillforts, burial mounds, standing stones, etc.) located in particular areas of a landscape (Madry and Crumley 1990; Gaffney and Stancic 1991; Wheatley 1995, 2004; Gillings and Goodrick 1996; Llobera 1996, 2003; van Leusen 1999). Indeed, for almost 30 years visibility has played a significant role in the debate about human-centred forms of archaeological spatial analyses. The importance of visibility studies in the archaeological discourse is further exemplified by a large number of dedicated chapters and sections in most of the textbooks on archaeological GIS edited in the last 20 years (Gillings and Wheatley 2002; Chapman 2006; Conolly and Lake 2006).

Despite the popularity of the method, to date, most GIS-based visibility analyses are performed with 2.5D terrain models used as spatial references for simulating the topography of the studied landscape. Most of the attempts to move towards a fully 3D approach (in which vertical visual obstacles are also included in the form of 3D models) did not involve the use of GIS but rather a combination of other tools, including 3D modelling software and game-engine platforms (Earl 2007; Paliou et al. 2011; Paliou 2011, 2014; Opitz 2017).

The following examples draw upon case studies in which a vector-based approach towards visibility analysis has been employed, with the analytical process entirely performed in a GIS environment. 3D vector-based visibility analysis has previously been employed in fields other than archaeology, such as urban planning and geography, with examples featuring the use of line-of-sight (LOS) for the analysis of urban spaces (Yang et al. 2007; Rød and van der Meer 2009; Moser et al. 2010; Suleiman et al. 2011).

The case studies presented in this chapter, with a main focus on the study of a Pompeian house and its visual configuration, describe an approach that has been made possible by the recent advances occurring in 3D GIS, where 3D surface/boundary models can now be imported and managed.

DOI: 10.4324/9781003034131-7

6.2 Investigating a Pompeian house

The space of a Pompeian house was a particularly interesting context for testing novel methods of 3D visibility analysis. It provided unprecedented opportunities for exploring the use of a GIS platform for managing and analysing 3D boundary models with a relatively high geometrical resolution. It allowed us to perform the analysis in a georeferenced space in which different spatial data sources could be integrated. On a general level, testing a 3D-based LOS was a way of investigating issues connected to human presence, perception, and movement inside an ancient space through a digitally informed approach, following formal rules that can now be extended and re-applied to other archaeological contexts.

It is now possible to define an innovative workflow that can make 3D GIS platforms effective tools in support of archaeological visibility analysis. The methodological approach proposed in this chapter aims at detecting visibility patterns so as to investigate human perception of well-defined categories of artefacts located in a virtually reconstructed space. The case study presented here relates to the Pompeian house of Caecilius Iucundus (Landeschi et al. 2016), where a wall painting and a graffito were examined and their visual impact on a hypothetical sample of observers was evaluated. These artefacts were chosen because of their explicit role as visual communication media. As is described in BOX 6.1, visibility analysis was performed based on a vector-based LOS approach in order to measure any direct visual connection between an observing point A and a target object B. The whole process was managed in a GIS environment, taking into account the role played by any visual obstacle that could have prevented a direct LOS connection between the observer and the observed object. The analysis made it possible to map the visual impact of the two examined objects throughout the space of the house by analysing, on one hand, the most favourable locations to gaze at the targets, and on the other, the portion of each target's surface that was most visually exposed.

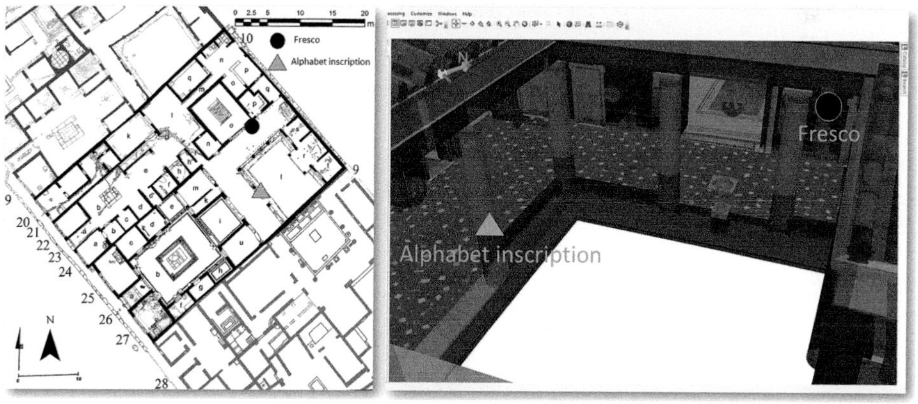

Figure 6.1 Left: plan of the house of Caecilius Iucundus (after Ezequiel Pinto-Guillaume), with the location of the examined artefacts, an alphabet inscription and a fresco depicting an erotic scene. Right: the same artefacts are located in their original x, y, z location and the model of the reconstructed space is used as a geometrical reference to perform the LOS analysis (Landeschi et al. 2016).

For this purpose, the analysis was conducted on two artefacts, a tiny alphabet wall inscription and a fresco depicting an erotic scene, located in different areas of the house of CI (Figure 6.1). Both of the artefacts were intentionally created to communicate visually with a number of observers/readers. Despite the limited amount of data included in the analysis (only two single objects), the importance of the experiment was that it illustrated the potential of overcoming some of the traditional limitations that existed in archaeological viewshed analysis (such as the lack of full three dimensionality or the underrepresentation of landscape elements). Additionally, this method provides new instruments for the analysis of Roman buildings and adds new tools and methods to the discussion about the public and private dimension in the study of the domestic space.

6.3 Digital reconstruction

This experiment was conducted as part of the SPP, initiated in 2000, with the purpose of recording and analysing a full city-block in Pompeii (Insula V, 1, Dell'Unto et al. 2015).

Among the different steps in the analysis, the reconstruction process of the house of CI played a prominent role. This is mainly due to the fact that the reliability of the simulation of human behaviours (movement, sensory perception) in any virtually reconstructed space heavily depends on the quality of the reconstruction itself. For this reason, it was crucial to develop this process following internationally acknowledged guidelines and formal languages for tracking the virtual interpretation process (London Charter 2006; Principle of Seville 2011) and a critical discussion involving specialists and project team members involved throughout the different phases of field documentation (for a detailed discussion on the method see also Barcelo et al. 2000; Forte 2010).

As has been extensively discussed (Demetrescu et al. 2016; Landeschi et al. 2016), the process of reconstruction was facilitated by the availability and accessibility of multiple sources, including legacy data (digitally scanned excavation plans and drawings, archaeological reports, etc.) and data collected in the field through the use of integrated acquisition techniques. Therefore, the southern part of the double atrium house of Caecilius Iucundus represented an ideal candidate for this analysis.

The examined elements were an alphabet inscription and a wall painting, which were investigated through vector-based LOS analysis. Both of the artefacts were targeted through vector lines of sight generated from a sample of observing locations. The main goal was to take advantage of a fully 3D approach, where vertical visual obstacles, represented by the boundary 3D models of the reconstructed house architecture, could be included in the analysis. To fully incorporate the multipatch geometry of the reconstructed house, a vector-based approach was then employed, being the only viable option at the time.

BOX 6.1 Line-of-Sight Analysis

The methodology described in this box is developed using ESRI ArcGIS Desktop/PRO version; the 3D models imported into the GIS were acquired using laser scanning; the point clouds were post-processed using Meshlab for generating the mesh and Blender for the texturing. The 3D reconstructions of the house were made using Blender (for a more thorough description of this method, see Landeschi et al. 2016).

The LOS tool, available through the 3D Analyst extension, allows users to build vector lines connecting an observation point (O) to a target point (T). O and T are both point shapefiles identified by x, y, z coordinates. Lines connecting O and T can be undisturbed or disturbed based on the presence of any visual obstacle, typically constituted by a portion of terrain (DTM) or a 3D model in the form of a multipatch.

For clarity in architectural spaces, LOS allows users to create horizontal and vertical maps featuring the visibility associated with wall surfaces (vertical) in relation to a grid of hypothetical observers evenly distributed in a room of the building (horizontal).

1 Setup of grids of observing and target points. Each grid can be customised depending on the height of the observers and the targeted object(s), as well as their density. For instance, in the Pompeian house, a density of 20 cm was chosen for the observation points (Landeschi et al. 2016).
2 Each grid point is the centre of a vector panel which describes its surrounding area so that the final assessment will show continuous horizontal and vertical surfaces indicating the frequency of visibility associated with the observers and the targets. Vector LOS is calculated through a dedicated algorithm and this connects observing and target point to each other. A descriptive attribute table defines whether each LOS line is a visible or non-visible one, depending on the presence of any possible visual obstacle between the connecting points.
3 The whole process, which typically consists of a huge number of points, is performed through a batch processing tool (ModelBuilder) that can be configured for building geoprocessing workflows including multiple parameters and tools.
4 Visibility grid maps are generated as a result of this batch processing.

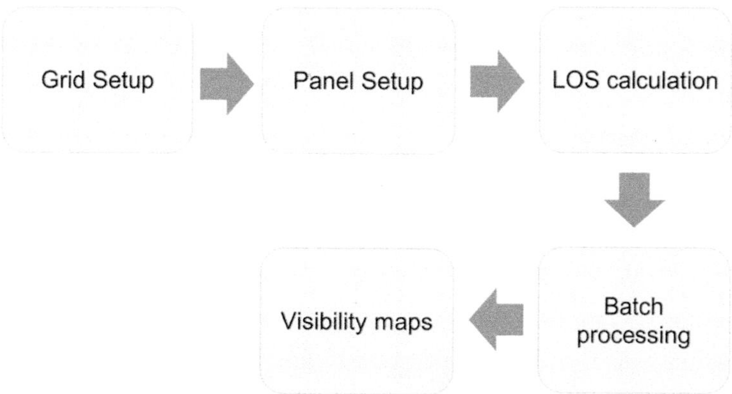

(*Continued*)

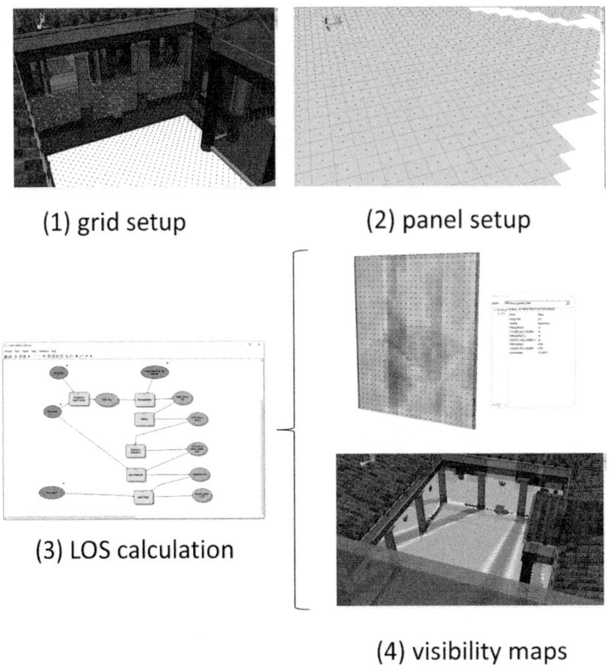

A few more notes: the described workflow can be performed either with ESRI ArcGIS Desktop 10.x or ArcGIS PRO. The ModelBuilder batch processing tool is available in both of the available releases and significantly speeds up the whole process. Further improvements in data processing might be achieved with high performing computing systems which include parallel processing tasks.

As per the description provided above (BOX 6.1), such an approach allowed us to measure the quantity of sight-lines impacting the whole surface of the target fresco and to obtain important information about the level of visual exposure associated with each spatial unit of the painting. This was crucial for understanding whether there was any particular pictorial element that was specifically intended for a higher display. Regarding the alphabet inscription, due to its small dimensions (9x2 cm), its surface was sampled in target units of 2 cm and the hypothetical observers limited to a buffering distance of 2 m from the inscription, as the small size of the writing would have prevented it from being read from afar. The reconstruction of its original location, a column in the peristyle, was based on the integration of several sources including the estimation of the height above the pavement floor, expected to be around 1.10, as the inscription was originally intended for children learning Latin (Landeschi et al. 2016).

Similarly, the original location of the wall painting was reconstructed and its surface was divided and sampled into target units to fully cover the whole surface of the fresco.

In this case, it was particularly interesting to evaluate the visibility of the fresco among the sample of hypothetical observers evenly distributed throughout the space of the house, with a sampling distance set to 20 cm and a height above the ground floor of 1.65 m. The LOS measurement was performed so as to produce a 3D polyline shapefile with an attribute table expressing two possible values (1,0), depending on the presence or the absence of any obstruction between fresco and observer.

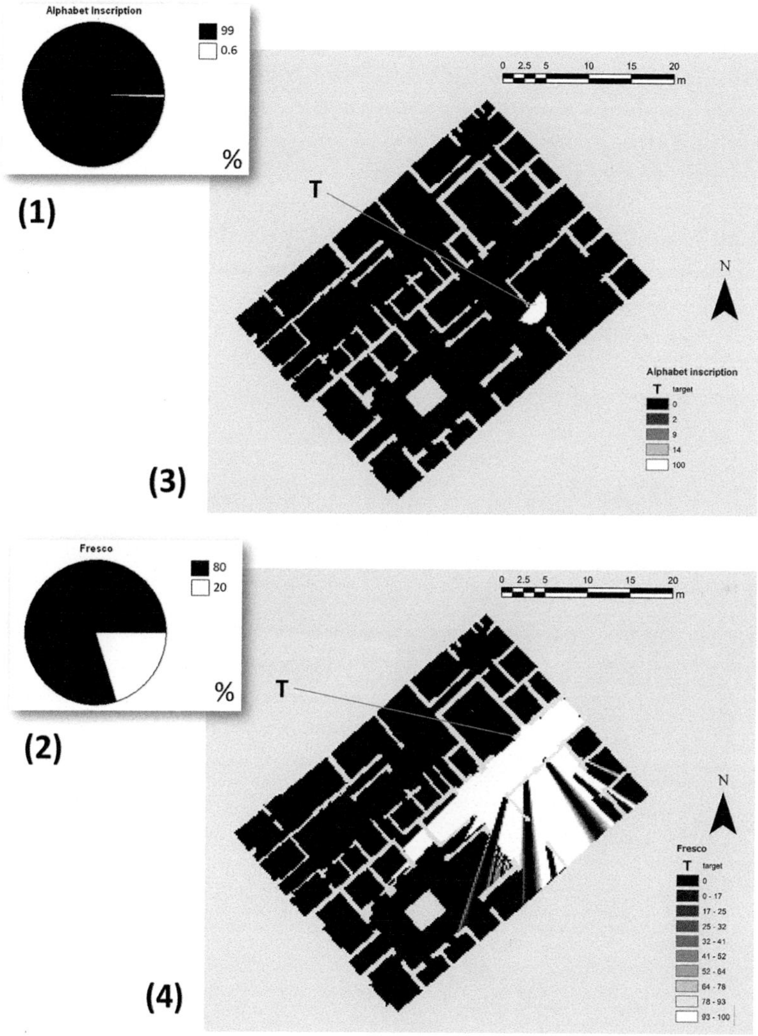

Figure 6.2 Comparative result of the 3D LOS analysis. Expectedly, the alphabet inscription's visibility is significantly lower than that of the wall painting. As graphs 1 and 2 show, only 0.6% of the overall house area is suitable for reading the inscription, while a much higher number of locations (20%) provided an undisturbed view of the fresco. Detailed, normalised frequency maps (3–4) show how much of the target object's surface is visible from each observing point.

6.4 Results

This assessment showed that a higher degree of visibility (i.e. locations from which one could have had a better view of the target object) was associated with the wall painting than with the alphabet inscription, as was expected (20% against 0.7%, Figure 6.2). In addition, the number of undisturbed sightlines connecting each observing point to the points spread across the target surface was calculated. Depending on the presence of vertical visual obstacles, the frequency value of undisturbed sightlines could have changed significantly. As a final result, the values of frequency were normalised according to the percentage values, with 100% visibility associated to those locations from which any observer would have had an undisturbed view of the entire target object.

Regarding the alphabet inscription, the reduced visibility value resulting from the analysis seems to confirm that this series of letters were originally inscribed for a reduced number of readers (children?) standing very close to the columns hosting the graffito.

As for the wall painting, when it comes to the analysis of the observers' frequency maps, it was possible to observe the predominance of the peristyle space as the most visually connected area of the house. As has been described elsewhere (Landeschi et al. 2016), the overall visibility of the wall painting was expectedly much higher than the alphabet inscription, and this was undoubtedly due to the different purposes that each

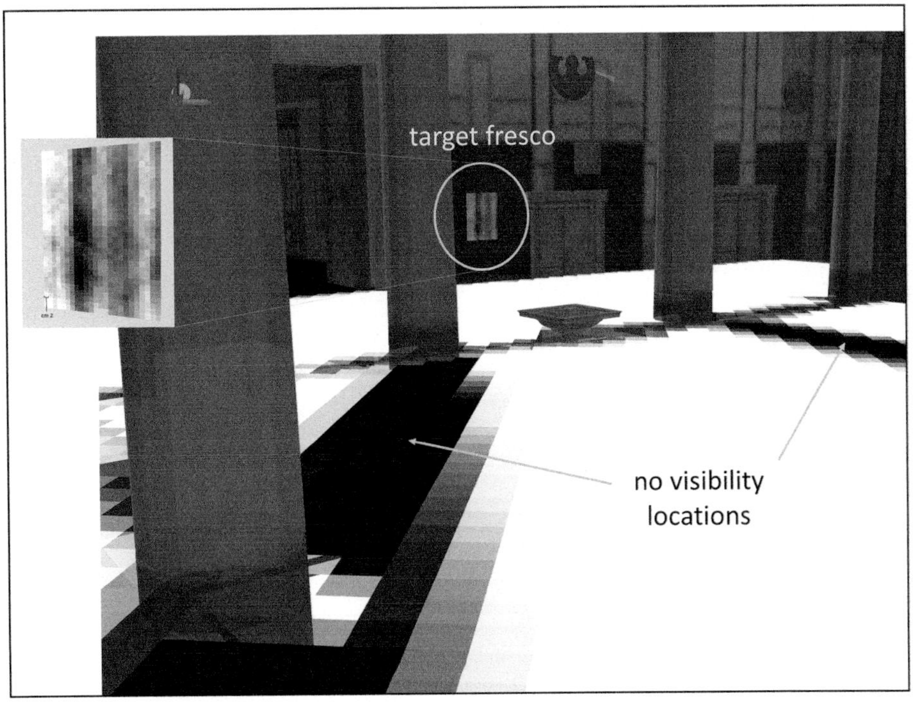

Figure 6.3 Visibility resulting from LOS analysis is clearly affected by the presence of vertical visual obstacles. By comparing frequency maps (of observers' locations and portions of visually exposed wall painting, respectively) it is possible to observe the role of the peristyle columns in obscuring views of the scene represented in the fresco.

artefact served. However, the actual visibility of the wall painting in its original context was likely to be lower than the results suggested by the LOS. This is because LOS analysis does not provide qualitative information about the visual acuity of the targeted object. Looking at the frequency map of the painting's surface, the analysis showed that there was a lower degree of visibility of some portions of the painting associated with the presence of columns in the peristyle garden which prevented a full view of the iconography (Figure 6.3). Otherwise, the painting's surface frequency map did not show any particular pattern as all the figures represented in the fresco had a similar degree of visibility. From an interpretative perspective, this means that the wall painting was intended to be read as a whole scene, rather than having different elements designed to be more or less visible at certain distances. Therefore, based on the simulation's results, this fresco was probably not a particularly "eye-catching" piece, as its overall visibility score was not as high as its location would have suggested.

6.5 Critical aspects

Despite the great advantage of performing visibility analysis in a fully 3D space, and the possibility of relying on an accurate and transparent reconstruction of the space as it was in antiquity, there are still some critical aspects that need to be considered, namely those that deal with the limitations of representing one person's visual attention in a GIS environment.

There are factors such as illumination, shadowing effects, and angle of view that significantly affect the visual experience of a user gazing at the wall (Ogburn 2006). Here, these factors would have made the painting iconography largely unintelligible for people entering the peristyle from the corridor. Having an undisturbed view of an object does not imply its visual acuity, nor the possibility for the human eye to "read" and understand artistic content (Bernardini et al. 2013). Depending on image size, observers can be prevented from gaining a full understanding of the whole scene represented in a fresco, and this issue must be carefully considered when applying formal methods, since the distance of observation can play an important role. Concerning the illumination, previous studies demonstrated the ways in which the effects of natural and artificial light can significantly affect gaze and visual attention of an individual moving in a built space (Earl et al. 2013). For this reason, a quick test for observing how the illumination parameters could have affected the actual visibility of the peristyle fresco was performed by using built-in tools available both in ArcGIS Arcscene and ArcGIS PRO release. This was a simple test to check how external, atmospheric and light parameters might affect the wall painting visibility. Through the scene properties viewer, it is possible to set some parameters including atmospheric lighting, shadow effects, and information about the natural radiating light based on the position of the sun in the sky (based on date and time). Quite obviously, the visibility of the fresco significantly changes depending on the time of day, and this aspect must be carefully considered whenever a GIS-based approach to visibility is performed (Figure 6.4).

Nonetheless, the strength of this experiment was that it illustrated the value of adopting formal methods of analysis applicable in virtually reconstructed ancient built spaces. Relationships between the location of wall inscriptions or paintings with specific areas of a built-up space, either a single house or a city neighbourhood, have been studied (Laurence 1995:68), with visibility approaches being used to further explore the connection between visual exposure of iconographic elements and pedestrian movement

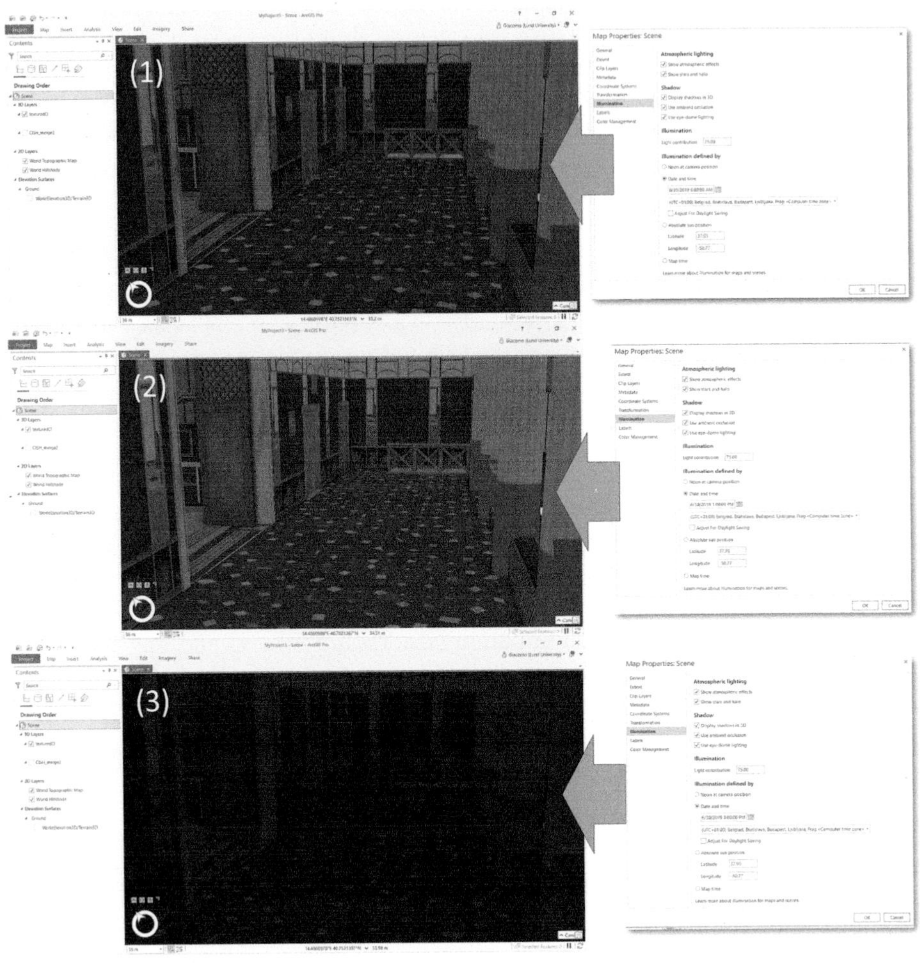

Figure 6.4 Scene properties can be customised to adjust sunlight conditions according to date and time. Three different scenarios in the summer season have been tested to highlight the significant contrast in visibility levels that can prevent or facilitate the iconography's observation and intelligibility. (1) Early morning around 6 am; (b) early afternoon, around 1 pm; (3) evening, around 8 pm. Based on the viewpoint of a hypothetical observer located at the corridor entrance (SW to the wall painting, 1.60 m above the ground floor), it is plausible that the iconography would not have been intelligible due to the distance and the poor illumination of the wall hosting the fresco.

(Paliou and Knight 2013). Using formal methods in a fully 3D environment allows the users to investigate visually connected spaces more thoroughly.

Through the overlap of multiple visibility maps related to different wall inscriptions, it could be possible, for example, to generate cumulative maps where each spatial unit defines the total degree of visual connectivity of that portion of space. That value can be considered a sort of proxy for understanding which areas were likely to be passage points for the people accessing the house, assuming, as it has been argued (Benefiel

2010), that there was a connection between the location of wall inscriptions and the movement of people. Mapping visibility through a 3D GIS-based approach can therefore improve our interpretations of those areas of visual engagement where potential readers tended to stay still and read the message on the wall.

In addition, GIS-based visibility analysis can be further compared with complementary techniques, such as space syntax analysis (Hillier and Hanson 1989), which can provide important clues for identifying patterns of presence linkable to social encounters inside a built-up space (Paliou and Knight 2013).

Concerning the house of CI, this case study opened up new ways of combining geometrically complex 3D models and GIS to develop a novel pipeline for visibility analysis. The use of built-in batch processing tools such as the ModelBuilder now allows users to handle a significant amount of data that can be employed to detect patterns of visual connectivity in unprecedented ways.

From a methodological point of view, the examined artefacts were useful for demonstrating how 3D LOS can empower the interpretative process of an ancient built space. The work pipeline described here can be extended to a broader dataset and applied to multiple classes of objects, including decorative and architectural elements.

Visibility analysis tools can also contribute to sensory studies of ancient space, and 3D GIS can now provide users with the possibility to draw on a wide range of solutions for exploring the visual structure of space in a wide range of ways (Llobera 2007). Still, it is crucial to tackle the typical issues related to the underrepresentation of certain spatial phenomena as they were observable in the past. As previously stated, visual acuity is an aspect of perception that is often neglected in archaeological visibility studies (Ogburn 2006; Frieman and Gillings 2007), and this lack of information can create a significant bias, as well as inaccuracies in the setup of the models used to perform the analysis, including an excess of arbitrary choices in the making of the 3D reconstruction and the misuse of illumination parameters.

6.6 Advancing the method

To cope with these issues, recent studies have made significant advances towards a more effective use of GIS-based visibility analysis. Visual acuity, namely the ability of the human eye to discern details in an image based on a minimal angle of resolution (MAR), was explored by Polig et al. (2020) in a case study about the iconography of an Orthodox church in Cyprus. They developed an innovative pipeline in which visual acuity played a key role in the assessment. It was important to define observers and observed objects in their contextual space and then to perform analyses that would allow visual acuity to be further examined in a 3D GIS, so as to eventually define the visual qualities of the church iconography.

Models of the church architecture and wall paintings were created through a combined approach of 3D acquisition techniques. Then, lighting conditions and human eyesight limitations were reproduced in order to collect information that could be fed into the GIS-based analysis of visibility. Interestingly, due to the excellent preservation conditions of the site, it was possible to perform the light surveys in situ, allowing them to determine the visual acuity values necessary to discern details in the scenes depicted in the wall paintings.

Different target groups of observers were chosen to conduct the tests in situ, and trigonometric calculations were used to determine the different visibility distances based on sources of light employed in the analysis (daylight, lightbulbs, candles), relying on experimental data from research in optics. As a final stage of this pipeline, the collected data were used to perform the analysis in 3D GIS. Within the imported 3D models of the church, a few key positions were chosen from which to perform LOS targeting to a set of previously digitised points on critical details of the paintings' iconography. The result of this analysis was then combined with the information previously collected about the range of distance within which a certain detail could be discerned, based also on the different light conditions at the time of the observation.

From an interpretative point of view, the combination of quantitative and qualitative data allowed researchers to better understand the functions associated with the depicted scenes and to establish what the role of the frescoes of St. John's Cathedral was, and how they were carefully planned and executed to build an important narrative that could be seen as "an illustrated book with a high didactic role" (Figure 6.5, Polig et al. 2020:15).

This work marks an important advance in GIS-based visibility studies, as the role of visual acuity and the effects played by elements such as illumination and angle of view are considered in the analytic process. Remarkably, this experiment also took into account different cultural backgrounds and the varying level of contextual knowledge of church visitors (expert vs non-expert), allowing the team to determine visual acuity thresholds to be added as additional parameters to the GIS-based assessment of the visibility of the church's main iconographic elements.

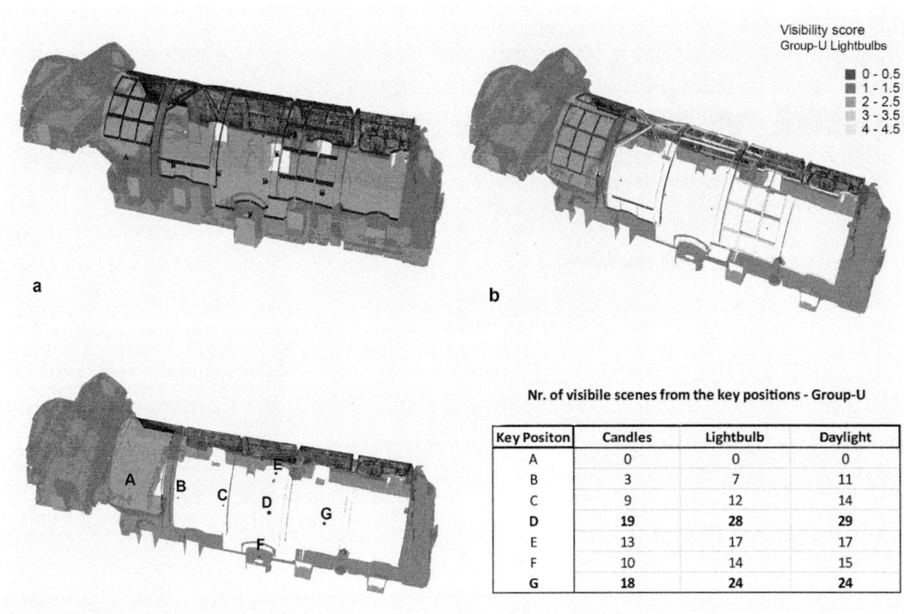

Nr. of visibile scenes from the key positions - Group-U

Key Positon	Candles	Lightbulb	Daylight
A	0	0	0
B	3	7	11
C	9	12	14
D	**19**	**28**	**29**
E	13	17	17
F	10	14	15
G	**18**	**24**	**24**

Figure 6.5 (a) 3D model of the church's nave visualised in GIS (ESRI ArcScene). (b) Painting scenes are marked by different colour gradients based on their visibility score (group-U with lightbulbs as a light source). (c) Key viewpoints marked by red points (A-G). (d) From each viewing position, the number of visible scenes is evaluated (image courtesy of Martina Polig, readapted after Polig et al. 2020).

Another aspect of LOS analysis that is worth exploring is the possibility of assessing the visual qualities of communal spaces, where ancient inhabitants moved and interacted. This could be a ritual space, a space for public activities, gatherings, processions or even domestic households. This is not a space associated and represented by a material object but rather an empty space that is delimited by well-defined boundaries (walls, fences, etc.). Representing this space and its related visual experience requires us to consider the use of volume information. This is an interesting aspect that so far has not been adequately dealt with. Indeed, a full-3D approach to visibility allows users to generate points of visibility that are described by independent x, y, z coordinates in a Cartesian space. This means that any portion of an empty volumetric space can be fractioned and sampled in a number of 3D target points that can be associated with values of disturbed/undisturbed view (Lundström 2017; Figure 6.6).

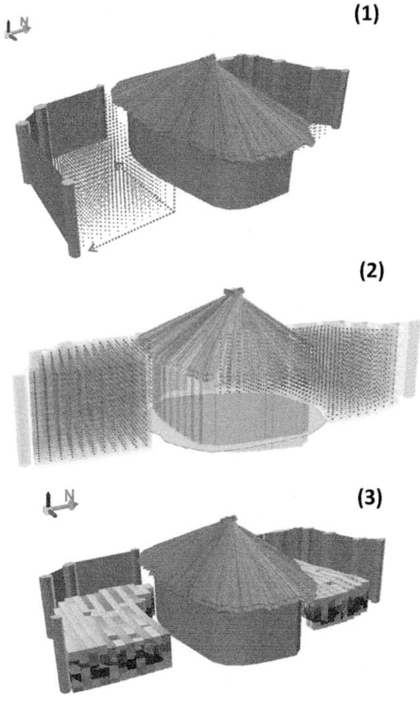

Figure 6.6 Volumetric visibility analysis can be used to assess a space for movement and encounters within its three dimensions (image courtesy of Victor Lundström, Arkeologerna). A space of ritual activity in a reconstructed mid-Neolithic cult house can be examined following a similar workflow: (1) 3D target points are digitised across the adjacent activity area of the cult house. (2) Each 3D point is turned into a centroid of a vector cube whose value of visibility is given by the sum of undisturbed sightlines connecting it to a number of observation points. (3) A volumetric frequency map displaying different degrees of visibility is generated and used to formulate hypotheses about the most visually exposed portion of space. (Lundström 2017).

6.7 Summary and conclusions

This chapter described some innovative approaches to visibility analysis that can now be performed in a full-3D environment in which GIS-based analytical tools are employed. In particular, LOS analysis has been discussed in relation to the visual qualities of inscriptions and wall paintings in a Pompeian house and, on a more general level, to define new workflows for investigating the spatial configuration of virtually reconstructed historical spaces.

More case studies were mentioned to generate new insights into the applicability of LOS analysis in very diverse research contexts. Medieval churches and prehistoric settlements, along with many other archaeological study foci, can now be re-examined by taking into account a human-centred form of investigation.

As will be further discussed in Chapter 8, more possibilities for the analysis of the individual visual experience will be provided by the combination of VR-based techniques of data collection with their integration in GIS. Eye-Tracking and Motion Capture technology, just to name a couple, can allow us to capture one person's visual attention in a more dynamic way by taking into account the individual mobility within a virtual environment and thus include the movement and the temporal dimension of the visual experience in an unprecedented way.

7 Volumes

7.1 Volumetric analysis in archaeology

In recent years, there has been an increasing body of literature presenting case studies in which volumetric information has been produced and employed in support of archaeological interpretation, specifically in the context of stratigraphic analysis (Gavryushkina 2021), historical architecture (Buccellati 2016) and bio-archaeology (Kazzazi and Kranioti 2017). By relying on different techniques and methods of data representation, including solid and surface/boundary models, all of these case studies have sought to address the main limitations stemming from traditional approaches to site/artefact documentation, and introduced work pipelines where the volumetric information was essential for the comparative study of material culture. In this context, although a GIS-oriented approach to volumetric analysis was envisioned already in the 1990s (Reilly 1991), there have so far been very few applications of volumetric data in GIS. This is due to the lack of workable pipelines for handling this data format, such as voxels or enclosed vector boundary models in a georeferenced space. To date, most of the available GIS software hasn't allowed users to handle these data in spatial relation with other objects (including raster and shapefiles), and this aspect still constitutes a major limitation to the applicability of volumetric forms of analyses.

A good example of volumetric data representation in a GIS environment was provided by an intra-site documentation project carried out at the site of Paliambela Kolindros, where a formal workflow for documenting the excavation in 3D was set up (Katsianis et al. 2008). One of the main advantages was the possibility of managing the whole process in a single software application and to customise it to the needs of the project. Compared to other data standards, 3D boundary representation here is introduced as an ideal option for coping with the geometrical irregularity that characterises stratigraphic units. 3D vector layers were created partly as a result of the digitisation of cartographic datasets, and partly as a result of the collection process in which single artefacts were assigned with 3D boundary models as symbols. Multipatch objects related to architectural features were mostly created from elevation TIN in an external program (SkecthUp), while stratigraphic units were obtained through a programming routine. Dedicated programming routines were also developed for performing basic calculation tasks that could allow archaeologists to retrieve information from this 3D database more efficiently and spatially relate each entity's position in the excavation space (Katsianis et al. 2008) (Figure 7.1).

DOI: 10.4324/9781003034131-8

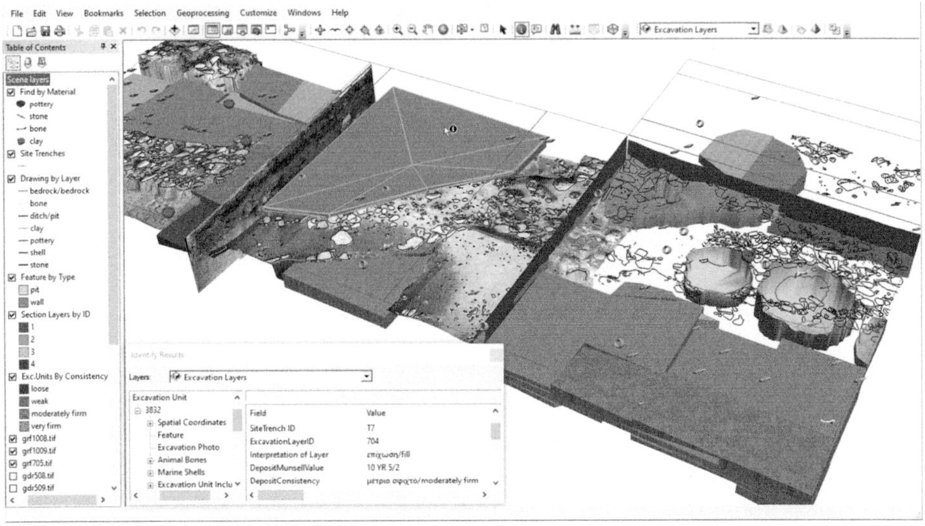

Figure 7.1 Trenches 5, 7, and 10 from the Paliambela Kolindros archaeological project (view from the Southeast). Trench outlines are depicted as dotted lines, excavation units by deposit consistency values as textured closed multipatch objects, finds by material using 3D symbols (boundary models), features by type as open multipatch objects, sections by stratigraphic layer as polygons, and drawings by layer type as lines. Plan and section photomosaics are imported as geotiff files and draped over 2.5 DTMs. The onscreen selection displays feature attributes and database relationships (*Image provided by M. Katsianis and the Paliambela Kolindros archaeological project*).

7.2 Analysing a prehistoric cave

While most of the existing approaches to volumetric data representation in archaeology have shown some innovative ways of displaying stratigraphic units (including the examples provided in Chapter 4, BOX 4.6) and, at a more advanced level, to query and highlight artefacts and unit distribution within the 3D space of a trench, there are as yet not many examples of how these data can be analysed in order to produce new informative layers that can be spatially related to the source material used to perform the analysis.

The approach proposed here draws upon a case study from a prehistoric cave in Gotland, and it is based on extruded boundary models generated as a result of an intersection between digitised excavation plans and a 3D-scanned model of the cave, used as a geometrical reference.

The hope here is to demonstrate the contribution of 3D GIS to put some historical excavation drawings (maps and sections) in relation with the original context by digitising them and extracting the spatial information required for virtually reconstructing the original stratigraphy as it was observed and excavated by archaeologists at the end of 19th century.

Such research has been carried out in the prehistoric cave of Stora Förvar, on the small island of Stora Karlsö, in south-eastern Sweden. This work was initiated as part of "The pioneer settlements of Gotland" research project, where combined methods of digital acquisition and analysis were tested with the aim of virtually reconstructing the stratigraphic sequence, originally excavated in 30 cm-thick arbitrary layers.

As a further step, all the available information from historical archives has been examined in order to spatially connect different classes of artefacts and ecofacts to their original context, granting the possibility of detecting some interesting patterns for defining which portion of the sequence corresponded to the Mesolithic occupation phases (Figure 7.2).

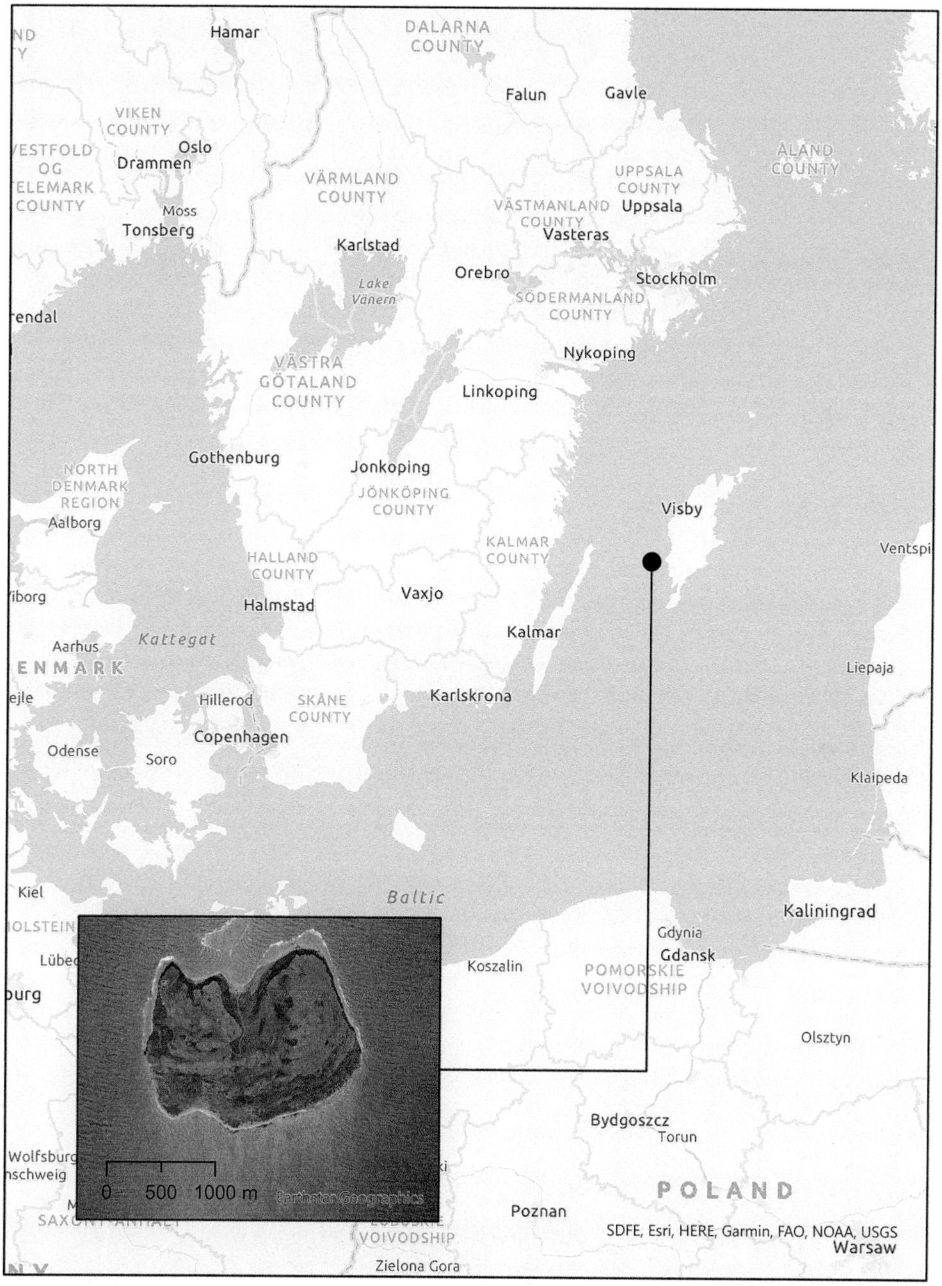

Figure 7.2 The location of the Stora Förvar prehistoric cave, on Stora Karlsö island, south-eastern Sweden.

7.2.1 Study context

Due to the particular nature of this research project, an approach based on the combination of 3D GIS and legacy data was essential. As Stora Förvar is a very prominent site in terms of available information, it provided unique opportunities for the study of the Stone Age in Gotland. Still, some major issues have always prevented archaeologists from performing a contextual analysis of the sequence of units and their related artefacts. The cave sequence was excavated between 1888 and 1893, and the final report was published 50 years later by archaeologists who were not directly involved in the original excavation. More data, including faunal remains, were collected by other archaeologists at different subsequent stages. Recently, in 2013, a new field campaign was conducted to investigate the state of preservation and integrity of the Mesolithic component in the still un-excavated portion of the stratigraphy (Apel and Stora 2017b).

To integrate information collected at very diverse time stages in a homogenous and consistent way, and to spatially locate artefacts and classes of ecofacts in their original arbitrary layer or spit unit, an innovative approach was developed with three aims: (a) recontextualising artefacts and ecofacts retrieved during the late 19th-century excavation in the stratigraphic sequence, (b) generating density maps in which the relationship between the volume of each unit and the number of retrieved objects per class is presented, and (c) increasing the knowledge about the Mesolithic habitation of the cave.

7.2.2 Method

As a first step of the process, it was important to rely on a 3D surface model of the cave that could be used as a geometrical reference for building the virtual sequence and for performing the expected analysis. For this purpose, the entire cave of Stora Förvar was digitally scanned in the summer of 2013 using a laser scanner (Landeschi et al. 2018). This method was particularly suited for the cave due to the light conditions that made any alternative approach, such as image-based 3D Modelling, very difficult.

To reconstruct the sequence as it was observed by the archaeologists at the end of 19th century, different source material had to be collected and combined in a geodatabase management system (GDBMS) following a relational structure as described in Figure 7.3 (ESRI ArcGIS Desktop release, 10.3). These data included source material accessible through the Antiquarian Topographical Archives, which consists of artefact lists, hand-made maps, and drawings on the site and field report. These legacy data were then digitised and integrated with "digitally born" source material derived from field data collection. Another important data source was the geographical data available through the web portal of the Swedish National Land Survey agency (Lantmäteriet 2015), which allowed the archaeologists to georeference all the previously described data and to add more informative layers to the geodatabase in the form of multispectral images and LIDAR-derived data.

7.2.3 Reconstructing the sequence

To reconstruct the sequence, different sources were brought together and a geodatabase management system was designed for including all of them. The main idea was to take

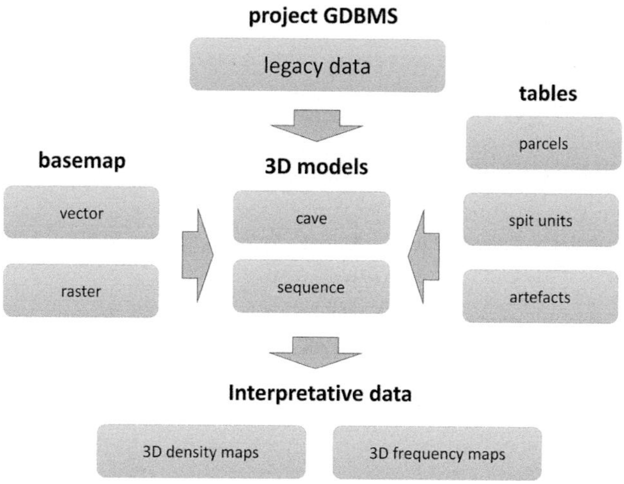

Figure 7.3 Workflow of data acquisition, implementation, and analysis, including different sources employed to feed the GDBMS. Interpretative data, in the form of 3D density and frequency maps, were obtained as a result of this integration.

advantage of the relational structure and topological rules so as to make it possible to contextualise finds, units and parcels in the 3D space of the cave, represented by the boundary 3D model resulting from the laser scanning acquisition.

To fulfil the needs of the project, datasets were organised according to their different formats and the original repositories they were derived from. The basemap consisted of orthophotos and DTMs that were used as a reference for placing the 3D model of the cave in relation to the landscape of Stora Karlsö. More detailed information regarding the original placement of units and parcels in the sequence excavated in the 19th century was provided by legacy data available through the historical archives and digitally scanned before the GDBMS implementation. Interestingly, most of these data consisted of 20th-century field documentation that was produced and published about 50 years after the completion of the excavation. Hand-made plans and profile drawings constituted an essential source of information for understanding how the cave sequence appeared just before the 19th-century excavation. The position of the vertical parcels and the horizontal arbitrary layers (spit units, see also Hughes and Lampert 1977) that composed the sequence was reconstructed based on a digitally scanned hand-made profile drawing that perfectly described the spatial placement of the sequence elements and their relation with the cave.

Non-spatial datasets were represented by tables containing information about the numbering system assigned to the parcels and units of the sequence, as well as the distribution of different classes of artefacts and ecofacts throughout these layers. A central element of this GDBMS were the 3D models of each single unit, created as multipatch feature classes (according to the workflow described in BOX 7.1), which eventually were used to assess the frequency and density of finds.

7.2.4 Data setup

Defining the architecture of the GDBMS is an essential step before proceeding with the data implementation. Then, the main spatial model to import (and that constituted the focus of this analysis) was the 3D surface model of the cave, which was derived from the laser scanning acquisition in the form of a 10-million-point cloud that was decimated and reduced to 1 million points, so as to be easier to handle in a GIS environment. More recently, following some technical improvement of the software, it would be possible to directly import the high-resolution model (see BOX 4.3). In terms of format, different choices could be made and a 3D surface/boundary model imported according to standards such as VRML or a Wavefront OBJ file. The conversion into a multipatch allowed us to manage the cave model in the GDBMS and connect it to the general database structure. Georeferencing information was provided by total station ground control points acquired on site. As the main purpose of the analysis was to use the cave model as a geometrical reference for reconstructing the original sequence and place all the units in their original context, no textures were added, which also made the process of visualisation less time and resource-consuming.

As a next step, the digitally scanned hand-made profile drawing was properly rescaled, imported as a multipatch feature class and placed in the right spatial relation with the model of the cave. Such a drawing was essential for identifying the position, thickness and length of each unit, as they were excavated based on an arbitrary method. The drawing was used as geometrical reference together with the digitally scanned handmade excavation plan of the cave, so as to be able to digitise unit and parcel boundaries marked with combinations of letters and numbers as 3D shapefiles. In this respect, the original sequence was divided into six different parcels that were marked with letters A to I, starting from the cave entrance to the innermost part.

A thorough description of this process is provided in BOX 7.1, representing a repeatable workflow that can be applied in different research contexts.

BOX 7.1 Building a sequence of volumetric extruded layers

The methodology described in this box is developed using ESRI ArcGIS Desktop/PRO version; the 3D models imported in GIS were acquired using Laser scanning; the point clouds were post-processed using Meshlab for generating the mesh. Raster images were imported and used as references for generating the volumetric information.

By taking the vertical profile drawing as a geometrical reference, each intersection between parcel and unit interface boundaries was marked and digitised with a 3D point shapefile (**1**).

A polyline shapefile was defined to mark the horizontal interface boundaries between each spit unit, connecting each intersection point (**2**).

Then, polylines marking each unit interface were converted into horizontal panels intersecting the cave model walls (**3**). By applying an intersecting calculation, it was possible to derive 3D polyline shapefiles that perfectly fitted the outermost boundaries of each spit unit (**4**).

The horizontal space enclosed within each polyline space was then converted into a 3D polygon shapefile (**5**). Basically, each spit unit was marked by an uppermost and a bottom polygonal plan defining its interfaces.

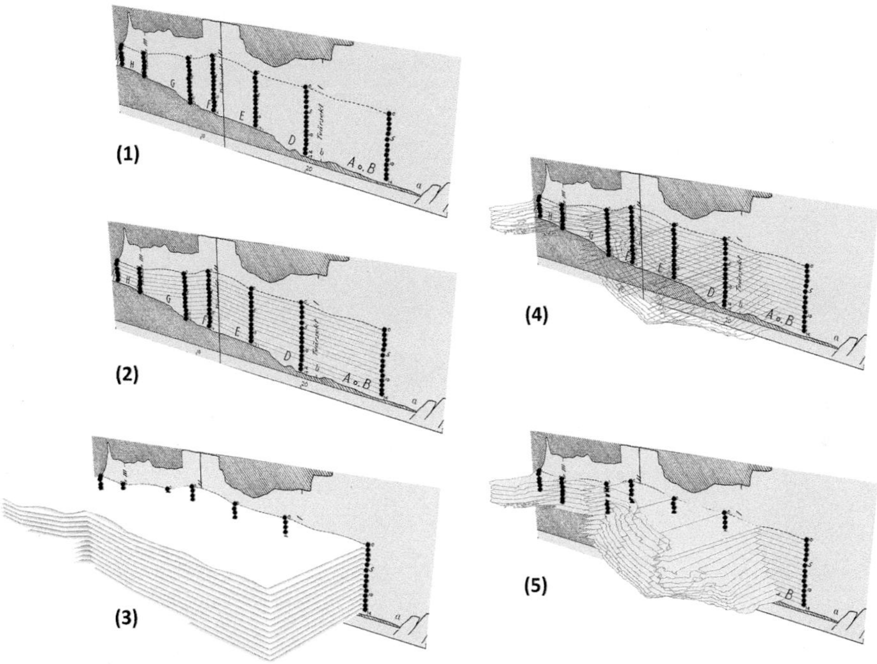

As a next step, it was necessary to increase the number of sub-layers within the 30 cm-thick spaces defined by each spit unit (**6**). In this way, it was possible to obtain a more "continuous" series of parallel polygons that could eventually be extruded to create a closed volumetric space for each unit. By creating sub-layers at a regular Z interval of 2 cm, each spit unit was filled with 15 separate overlapping plans (**7**). Creating these thinner sub-layers meant that the team could minimise the approximation error from the vertical extrusion process, which does not consider the surface irregularity typical of cave walls.

Concerning the external part of the sequence that was protruding outside the cave and that could not have been defined based on the intersection with the cave model's walls, it was possible to determine its length, width and depth based on the combined information derived from the profile drawing and the original excavation plan.

The external portion of the sequence was reconstructed based on the geometrical boundaries described by the original excavation drawings. (**8**) Width was reconstructed based on the excavation plan. (**9**) Length and depth were modelled according to the boundaries traced on the profile drawing.

(*Continued*)

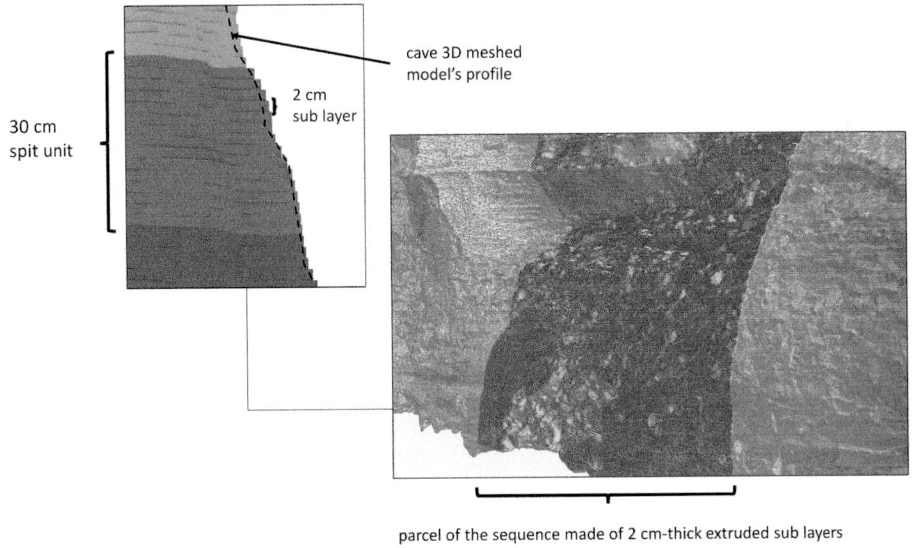

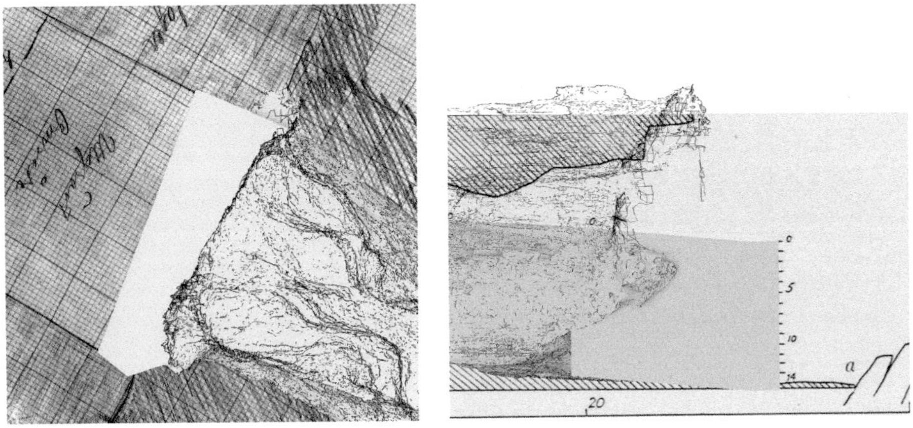

(1) (2)

A few notes: to speed up this workflow, there are several built-in solutions in ESRI ArcGIS to run most of these tools in batch-processing mode. ModelBuilder is a pretty easy-to-use visual programming tool to fulfil this purpose and let the user create customised workflows of tasks to be performed in sequential order.

7.2.5 Results and discussion

As a final result, each spit unit was composed of 15 separate 2 cm-thick volumetric sub-layers. In total, more than 7,000 single volumetric units were produced to fill in, as a composite volumetric model, the space of the cave occupied by the sequence recorded in the late 19th century.

Through table join operations, all the sub-layers were connected and grouped based on their spit unit number. Then each unit was linked to the artefact geodatabase tables in order to make the volumetric sequence a queryable space to search for the position and distribution of the different classes of finds recorded in the historical archives but never really spatially located. This represented a crucial step for the analysis of the collected data, as it allowed specialists to investigate the original distribution of artefacts and ecofacts and infer possible patterns of presence associated with the different chronological phases of habitation of the cave.

Remarkably, reconstructing the unit layers as enclosed multipatch feature classes made it possible to calculate the volume value for each spit unit. In this way, 3D maps of find distribution were produced, allowing us to calculate the frequency and density of the different classes of find. For the purpose of this study, the distribution of pottery, harp seal bones, and bone harpoons were considered important indicators of the relative chronology of the cave, and therefore the results of their analysis were discussed (Figure 7.4).

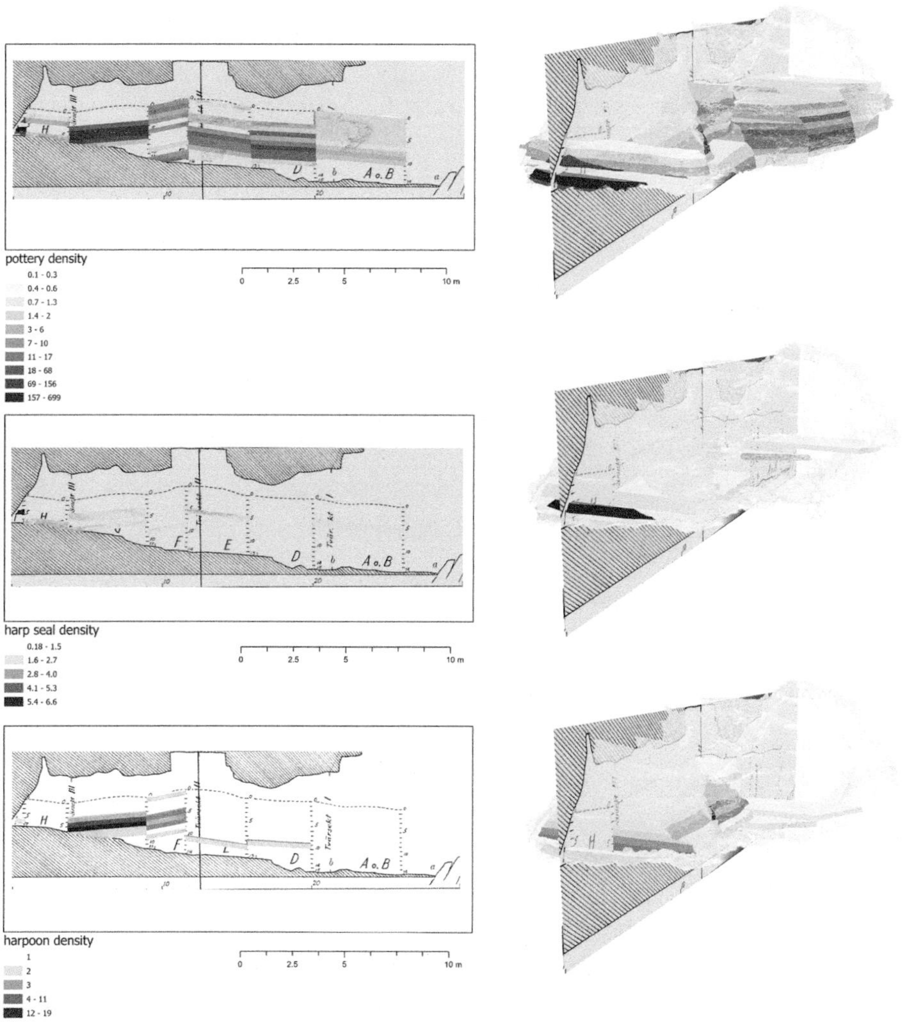

Figure 7.4 Density maps obtained by examining finds distribution throughout the sequence units. Three classes of artefacts/ecofacts (pottery, harp seal bones, harpoons) are particularly useful as relative chronology indicators.

7.2.5.1 Pottery analysis

Regarding the examination of pottery fragments within the Stora Förvar sequence, the spatial distribution analysis allowed us to gather some interesting conclusions. An overall trend of diminishing pottery density was observed, with very few fragments registered at the lower levels of the sequence. Based on the combined analysis of pottery density and osteological analysis, it is possible to argue that more spit units than previously expected could be dated to the early Mesolithic component, and further investigated with complementary methods including radiocarbon dating and artefact analysis.

Such a conclusion is drawn because the use of ceramic technology is a chronological indicator for the early Mesolithic portion of the cave, to be connected with a production

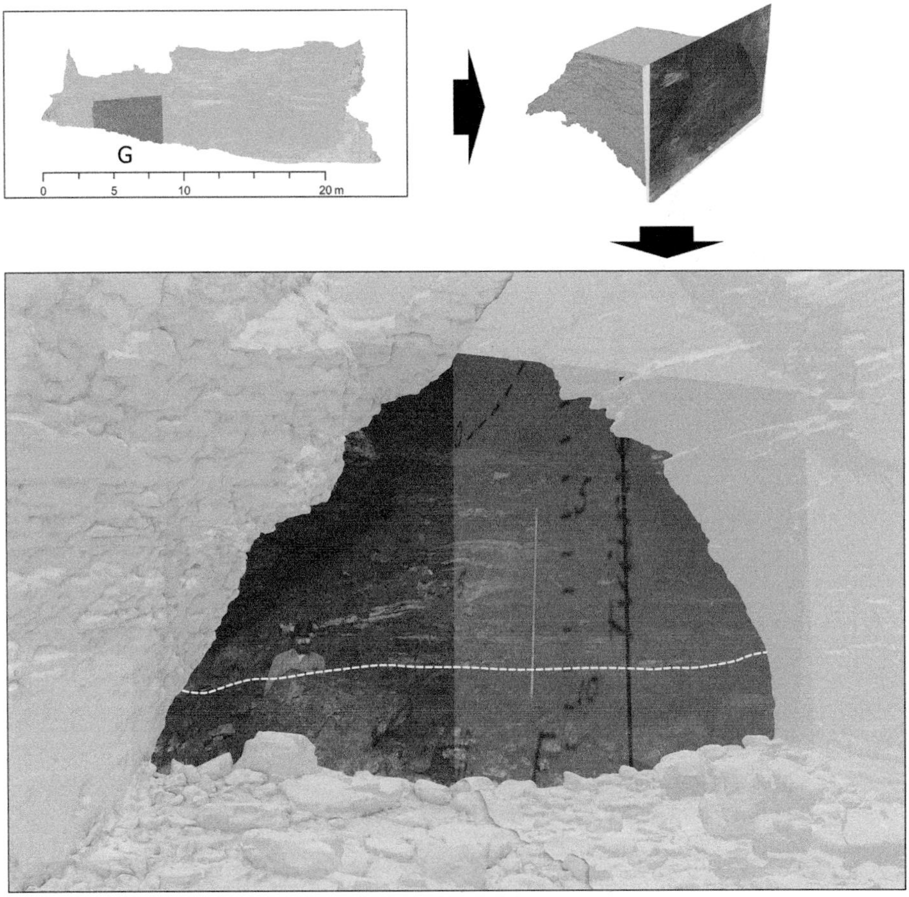

Figure 7.5 Georeferenced historical photograph showing the front section of parcel G. Dashed line indicates the change in the character of the stratigraphy which may well correspond to the hiatus, when the cave was left unused for around 1,000 years. In the georeferenced space it is possible to compare its position with the units' location as per the description provided by the original profile drawing. Reproduced with permission of the Antiquarian Topographical Archives, Stockholm (Photo 1492:2) (Landeschi et al. 2019).

operated by pioneering groups of hunter-gatherers preceding the Kongemose and Ertebølle cultures (Landeschi et al. 2018).

To sum up, pottery and other categories of finds, including harp seal bones and harpoons, are noteworthy indicators of the relative chronology of Stora Förvar sequence. Their spatial contextualisation, together with the analysis of absolute chronology indicators, is crucial in several ways for defining the original space of the stratigraphy corresponding to the Mesolithic phases of habitation.

The 3D nature of the analytic environment made it possible to reconstruct spatial relationships among the different indicators so as to draw more solid conclusions about the stratigraphic boundaries associated with the Mesolithic and post-Mesolithic events.

Interestingly, both pottery and faunal density values provided an important terminus to be associated with volume quantities corresponding to these chronological phases, and their progressive increase throughout the upper levels of the sequence allowed us to more precisely locate the portion of the sequence under investigation. An additional indicator of the relative chronology (to be used as a terminus post quem) was represented by the harp seal bones, as this species was not present in Baltic basin before 6,000 Cal BP.

Additional clues to the spatial location of the boundaries of the Mesolithic phases in the sequence are represented by historical photographs from the 19th-century excavation. By looking at a historical picture of the section profile of parcel G, it was possible to recognise soil matrices of different colours that indicate a hiatus in the stratigraphy, corresponding to the uppermost boundary of the Mesolithic portion of the sequence, possibly indicating a period of abandonment. Georeferencing this photograph allowed us to place the observed hiatus in the right spatial relationship with the boundaries marked by the volumetric reconstructed sequence, the digitised hand-made profile drawing, and the cave model itself.

As for the absolute chronology, an important indicator was represented by the C14 dated faunal remains sampled from the wall concretions, which were detected in the innermost part of the cave. By georeferencing these samples it was possible to further refine the location of the chronological boundaries associated with the Mesolithic portion of the sequence by introducing an independent chronological indicator (Figure 7.5, Landeschi et al. 2019).

7.2.6 Final remarks

One of the aims of this study was to improve the archaeological interpretation of a Prehistoric sequence by taking advantage of an integrated approach of acquisition techniques and data sources, contextualised in order to simulate the original 3D relationships between recorded stratigraphical units and finds. A crucial element for such an analysis was the integration of legacy data and their contextualisation in the digitally reconstructed cave where the excavation was carried out. Three-dimensionality and volumetric information allowed us to re-analyse the available dataset quantitatively and assess the spatial distribution of multiple classes of artefacts and ecofacts, which produced patterns of cave inhabitation. In particular, three find categories acted as important chronological markers for identifying which units in the sequence belonged to the Mesolithic phases of inhabitation. Frequency and density values were thus examined and the resulting patterns were related to other chronological indicators, such as the sudden change of colour in the soil matrix visible in that portion of the sequence, illustrated by a historical picture of a section of parcel G and geolocated in GIS. Information about the relative chronology of the sequence was then placed in relation to the absolute chronological indicators that included

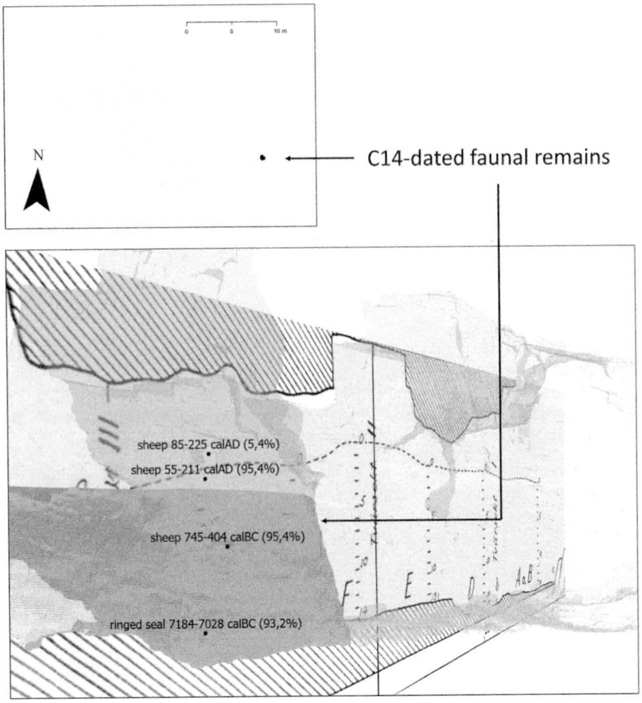

Figure 7.6 C14-dated faunal samples were digitally located in their original location on a portion of the wall in the innermost part of the cave. These independent chronological indicators constituted a crucial element for dating the sequence and its 3D positioning allowed us to better relate it to the location of the spit units.

C14 dated faunal remains recorded in situ from wall concretions after they were georeferenced and spatially connected with the previously mentioned datasets (Figure 7.6).

To sum up, despite some bias due to the inaccuracy of data recording and some level of uncertainty in the process of digital reconstruction of the cave, the work conducted in the cave of Stora Förvar demonstrates the importance of examining and interpreting archaeological stratigraphic data in a 3D context (Landeschi et al. 2018, 2019). A 3D-informed approach to spatial analysis allowed us to combine relative and absolute chronological indicators in a georeferenced space and to create new informative layers, based on the thorough examination of digital datasets and information provided by historical archives, such as excavation diaries and hand-made drawings.

The detected patterns allowed us to increase our understanding both of the site itself and of the pioneer settlement of Gotland. Notably, the role of legacy data introduces us to new possibilities offered by data re-use, which include the opportunity to contextually re-analyse material that was never studied in spatial relation to the context where it had been retrieved.

This kind of workflow can potentially be extended so as to re-examine field documentation produced in a "pre-digital" era and use it in support of site interpretation. This would allow more specialists to revise previous hypotheses by taking into account the spatial dimension in which different categories of finds and units were documented. By adding information about verticality and volume it would be possible to integrate more information about the original nature of the archaeological assemblage and relate these quantitative data with a geodatabase-structured environment where other specialists' data can be integrated. This would significantly improve a reflexive approach to site interpretation, where each single element collected can be revised and studied as a relational object whose significance is further enhanced by its contextualisation in a multiscalar and multitemporal perspective.

7.3 Handling geophysical data

As has been described in Chapter 3, one of the main limitations of geophysical data handling in GIS is the lack of well-defined pipelines for importing, visualising, and analysing volumetric datasets so as to put them in relation with database entries. Nevertheless, this is something that seems to be about to change, as there are now viable ways of managing these types of data sources. Interestingly, when it comes to GPR acquisition (for more information on this prospecting method, see Goodman and Piro, 2013), one of the main advantages is the possibility of collecting subsurface information at different depths and producing a data output in the form of a three-dimensionally interpreted anomaly. Such an object may very likely correspond to the archaeological feature under investigation, which can appear either as a void or as a "solid" structure.

By using dedicated software it is possible to export the interpreted anomaly both as a single 3D object in the form of a boundary model and as an ascii file that can be, in turn, transformed into a 3D grid of sampling points. Vector cubes can eventually be constructed from each xyz point (serving as a centroid), following the steps described in BOX 7.2.

BOX 7.2 Deriving a 3D grid of vector cubes from an ascii file

The methodology described in this box is developed using ESRI ArcGIS Desktop/PRO version; the 3D points imported into the GIS were acquired using Ground Penetrating Radar (GPR); the point clouds were post-processed using the 3D Analyst extension and visualised in ESRI ArcScene.

While this pipeline would be typically applied to data derived from Ground Penetrating Radar (GPR) acquisition, it could be extended to any dataset resulting from a sampling of 3D points with associated attribute values (geological coring, fluid dynamics, light simulation, etc.).

Most of the software packages commonly used for processing data collected in these kinds of prospections allow users to generate and visualise 3D point clouds defined by x, y, z coordinates along with attribute values corresponding to the amplitude of the signal captured by the sensor. These datasets can be exported as an ascii human readable file which can be imported and opened by any GIS programme.

(*Continued*)

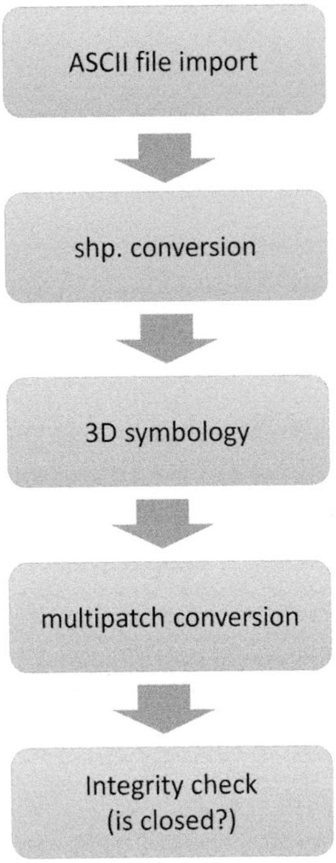

As a first step (**1**), the ascii dataset is imported and visualised as a spatial layer by using a dedicated geoprocessing tool (e.g. "Make x, y event").

The layer is exported and saved as a shapefile (or feature class). Through the 3D Analyst extension it is possible to visualise the dataset as a 3D point cloud in which each point is defined by its original x, y, z coordinates (**2**).

Under "symbology" it is possible to assign a 3D symbol to each point. This operation allows users to generate a dataset of vector cubes whose edge size is determined by the sampling distance separating each point in the 3D space. As a result (**3**), a seamless volumetric space will be defined through topologically connected cubes representing the interpolated value of each single sampling point.

In order to calculate volumetric values for each cube, it is necessary to export the 3D shapefile as a multipatch feature class and to run a geoprocessing tool to check whether the obtained geometry represents a "closed space".

To properly georeference the volumetric dataset (in case the raw data come in local coordinates) it is useful to apply a global shift function, available in third-party software such as Cloud Compare. In this way, the 3D point cloud can be imported into GIS and placed in its correct spatial location. If needed, it is possible to perform further adjustments in editing mode and roto-translate the dataset to better fit other geospatial data, as is shown in the figure below, where the interpreted archaeological electromagnetic anomaly (top right) was georeferenced according to the position of the ground anomaly (top left) previously identified in the Digital Surface Model and used as a spatial reference.

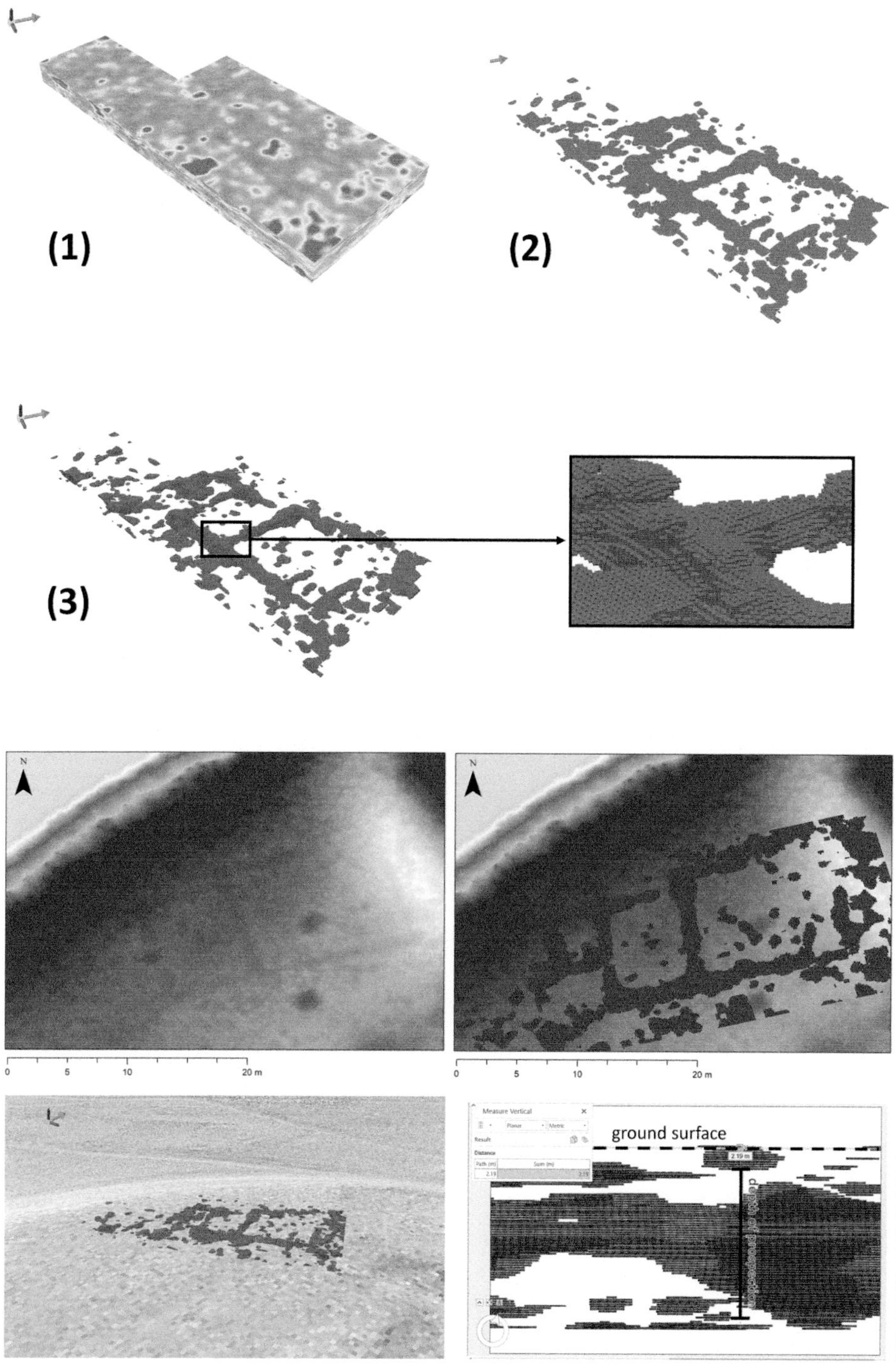

(Continued)

This operation allows users to handle the identified GPR subsurface anomaly in its full three-dimensionality and in the right spatial relation with the model of the terrain (bottom left). In terms of excavation planning, specialists can easily examine the depth at which the electromagnetic anomaly is met (bottom right) for each single unit of investigated land, or the amount of soil that needs to be removed before reaching the archaeological layers.

A few notes: This workflow can be followed either with ArcGIS Desktop 10.x.x or with ArcGIS PRO x. To optimise and adjust coordinate information contained in the ascii file, users should consider the possibility of using a 3D point cloud editing software before handling the data in GIS.

Volumetric information constitutes an important component of any approach to 3D spatial analysis. This chapter has introduced and discussed possibilities and limitations connected to the use of volumetric data in different research contexts. An important starting point is an understanding of which standards and formats are currently available for handling volumetric information. For this reason, a quick review of the main solutions currently available is presented and discussed in relation to relevant archaeological case studies. Archaeological excavations and prehistoric caves are particularly noteworthy due to the very peculiar nature of these contexts, where archaeologists need to face the geometrical "complexity" of the units they want to represent. In this sense, a special focus is put on the case study of the prehistoric cave of Stora Förvar, where an innovative workflow for the collection and the analysis of volumetric data is presented. Here, original excavation drawings were digitised and used as a geometrical reference for reconstructing the stratigraphic sequence of the cave and the volumetric information connected to each unit. Different categories of finds were subsequently connected to their original unit, and this made it possible to generate 3D density maps, enabling archaeologists to better understand the original patterns of distributions associated with the different classes of artefacts and ecofacts retrieved during the original excavation. As a concluding part of the chapter, a dedicated descriptive box introduces a workable pipeline for handling data derived from a geophysical prospection, where ascii files are imported and converted into multipatch feature classes in a 3D GIS, to be visualised as volumetric vector cubes.

8 Future developments

8.1 New sensors, new data, new perspectives

We expect future 3D GIS platforms to develop in multiple directions, including new analytical and editing tools for the analysis and manipulation of spatial data, and the implementation of Artificial Intelligence for supporting classification and analysis of large datasets of 3D information.

The introduction of new, game-changing technology such as multispectral sensors, Lidar, UASs, Eye Tracking (ET) and immersive technology is triggering an experimentation phase that should guide us towards the identification of new, unexplored research directions. 3D GIS practitioners have already started including these new systems in their methodologies, revealing the great advantages in merging these new datasets within the frame of archaeological investigation and spatial analysis.

8.1.1 Remote sensors and remote devices

Since the 1970s, satellite images have provided archaeologists with the possibility to analyse vast portions of the landscape through the use of different sensors (Lasaponara and Masini 2012). By using GIS for combining such images with other types of information such as Digital Terrain or Elevation models, it has become possible to obtain a broader and more clear overview of the different evidence scattered across the landscape, increasing the chances of identifying new patterns.

Despite the opportunities offered by GIS technology, the low resolution of satellite imagery, the impossibility of using 3D geometrically high-resolution surface models and the limitations of computers in managing high-resolution data represented a strong hindrance for the further development of these approaches in support of landscape studies. A revolutionary achievement for 3D GIS occurred with the introduction of airborne active sensors (Lidar) considered by many scholars working with remote sensing as the most impactful technology since the invention of photography (for the potential of this technology, see Opitz and Cowley 2013) (Bewley 2005).

Nowadays, Lidar technology constitutes one of the main information sources for recording landscape surfaces (particularly under forest canopies), and the increased availability of Lidar data within the framework of several disciplines encouraged the creation of dedicated tools for the management, use and analysis of these types of data within the 3D GIS environment.

Using 3D GIS for visualising Lidar datasets, maps, 3D surface, boundary models and 3D polylines and points, all in the same virtual space, represents an important asset

DOI: 10.4324/9781003034131-9

in heritage studies and provides researchers with the possibility of combining datasets recorded at different scales of resolution, thus achieving a broader representation of archaeological space (Figure 8.1).

The range of resolutions that characterise 3D spatial data often prevent scholars from performing more accurate observations, and the recent development and spread of remote controlled technology – such as Unmanned Aerial Systems (UASs) – has allowed us to use passive and active sensors such as laser scanning or digital cameras for the generation of spatial datasets with higher levels of resolution and coverage (Campana 2017). Specifically, the combination of UASs with high-resolution cameras or Lidar sensors for generating 3D dense models of large archaeological environments has resulted in far more efficient and cost-effective methods for mapping archaeological sites and features. The possibility of using such devices for recording the ongoing investigation has had a strong impact on archaeological field practice. Archaeologists are increasingly combining UASs and image-based 3D Technology (together with other sensors) for the recording of ongoing field campaigns (Waagen 2019), bridging archaeological records from micro to macro (Figure 8.2) and expanding archaeological practice beyond any previous expectation.

UASs allow archaeologists to gather data at almost any time of the day and year, customising the documentation and data acquisition process according to the immediate needs of the project. High-resolution multispectral data capture from UAS can be combined with high-resolution 3D representations of site morphology, providing important support for spatial analysis. The development of such an approach has permitted quantifying the location of archaeological assemblages scattered across a large portion of landscape, and has thus provided the opportunity to identify large-scale archaeological

Figure 8.1 A 3D model visualised in a 3D GIS platform, created using image-based 3D Modelling techniques, in spatial relation with a Lidar dataset provided by Lantmäteriet (under Lund University license).

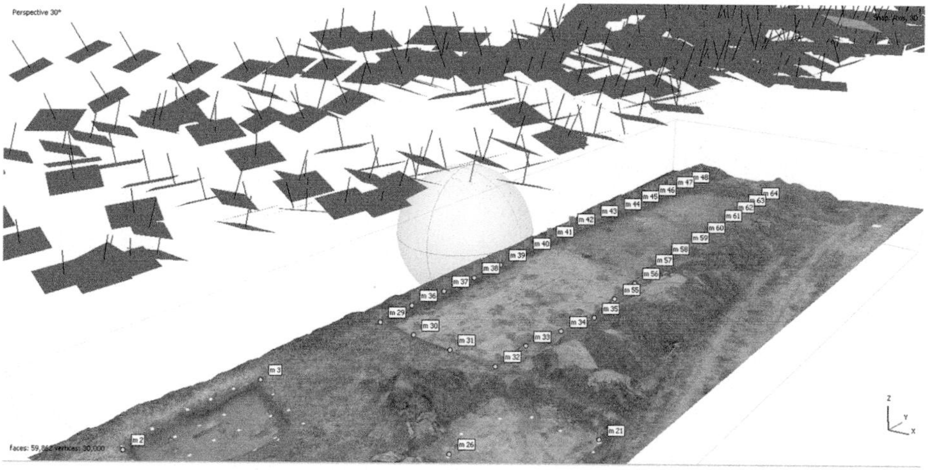

Figure 8.2 3D visualisation of the archaeological excavation of Vång, South Sweden. The trench was recorded using images acquired from a drone. The photographic dataset was then processed using Metashape 1.6 and imported into ArcGIS PRO. Courtesy of Vång project.

phenomena. The use of 3D Geographical Information Systems for the visualisation of such data allows us to review these results in a multidimensional space and spatial relation with other types of datasets.

The introduction of such new techniques and sensors, and the availability of large datasets of information, has promoted the development of new methods for the automatic detection of ground anomalies. Already at the beginning of 2000, high-resolution multispectral and panchromatic satellite imagery was used for automatically detecting archaeological features (Trier et al. 2009; Grøn et al. 2011). Similar approaches were then tested using raster images derived from lidar datasets (Trier et al. 2015, 2018) for identifying information using geometrical data visible on the digital terrain models (DEM/DTM).

More recently, this approach was extended to 3D points clouds (e.g. Lidar data). The possibility of grouping points into subsets or "segments" characterised by similar properties promoted a more dynamic use of these datasets for solving specific tasks (Matrone et al. 2020). By providing attribute table fields for specific groups of 3D points, it was possible to extrapolate information for supporting research across different disciplines such as architecture, civil engineering or landscape studies (Grilli and Remondino 2019). Recently, 3D GIS platforms started to include various classification methods for analysing remote sensing imagery data. A few software packages already include machine learning classification, convolutional neural networks and deep learning models that can be used to detect and classify archaeological features. The application of tools for the segmentation of point clouds or 3D models represents a possible future scenario which, if combined with 3D GIS, will certainly have a strong impact within archaeological studies.

8.2 Space, movement and sensory perception

8.2.1 GIS and the role of senses

Within the study of past spatial phenomena, non-material aspects of the ancient world play an increasingly relevant role. Consequently, sensory studies represent a very promising area of application for 3D GIS. So far, there have already been several publications dealing with the problem of investigating ancient space through a sensory informed approach (Fahlander and Kjellström 2010; Hamilakis 2014; Pellini 2015; Skeates and Day 2019). But although all of the published works deal with the problem of understanding the social construction of ancient space by employing a phenomenological perspective, it is very seldom the case that a viable workflow of integrated digital methods is proposed as a possible solution for capturing this ephemeral dimension (Landeschi and Betts forthcoming).

GIS have long been employed for investigating non-material aspects of ancient societies, with visibility studies playing a pivotal role in this context, as has described above. Still, by taking a 3D approach it would be possible to increase the dynamic nature of the simulated scenarios. This is an important aspect to consider, as there is a significant amount of complexity to deal with when tackling digital representation and the modelling of sensory perception.

To cope with this issue, integrated approaches to digital methods are required. Increased computational power, high-performance computing, and recently introduced active and passive sensors provide new opportunities for simulating interesting scenarios. Data can be collected with a high degree of resolution and virtually any aspect of human perception can now be captured and digitally represented: besides visual experience, it is possible to capture acoustic waves, use haptic devices to simulate tactile experience or map human movement in real-time. On a deeper level, it is even possible to use EEG to monitor brain activity and investigate aspects of low- and high-level cognition that can, in turn, foster our interpretation of a virtually reconstructed ancient space.

The unique nature of GIS enables users to map any form of spatially contextualised data in a consistent and coherent way. It is quite clear that human experience of space is a synaesthetic endeavour, and as such, requires us to face the problem of sensory stimulation in a controlled environment. As several authors have pointed out (Lock et al 2014), the experience of viewing prehistoric monuments involves movement (of the eyes and the body), and both seeing and moving should therefore be carefully examined. Dramatic advances in data-tracking technology (ET, motion capture, etc.) now make it possible to take advantage of more sophisticated sensors and formal methods of data capturing, allowing specialists to export these data in a format that can be studied in a GIS environment.

8.2.2 Visuality: new opportunities for research

As discussed in Chapter 6, the field of visibility studies has developed an increasing awareness of the role of visual acuity and the need for new instruments for examining human behaviour in relation to visual stimuli. It is now time to rethink the role of VR-based reconstructions, illumination effects, and human interaction with immersive environments. This is a necessary step if we wish to better understand the possibilities and limitations of the data we collect and learn more about the visual qualities of ancient spaces.

Although the use of immersive environments providing users with a "realistic" view of a pre(historic) space is not a new idea, there are almost no available examples of the application of analytical tools to explore the visual interaction between user and surrounding space. Among these tools, ET technology constitutes a valuable tool for capturing a user's gaze and formally assessing the time spent staring at a certain object, the order of visualisation and their movement within the immersive space. All of these aspects can dramatically improve the quality of any visibility-related experiment, especially when it comes to the study of ancient buildings or single sites (Figure 8.3).

The additional value provided by GIS-based analysis is the possibility of importing the data collected during a VR-based ET experiment and mapping human visual attention directly onto the 3D models. An iconographic motif in a virtually reconstructed medieval church, for example, can be displayed in a 3D GIS and the paths of movement of the user can be exported along with his/her gaze and visual attention over the different portions of the examined space. By integrating the results from multiple user sessions, it is possible to compile the measurements from the virtual environment and generate cumulative 3D maps of visual attention, that can then be used to define focal points, on a wall painting surface for instance.

This is possible because of the raster and vector-oriented analytic tools that allow users to perform several tasks such as map algebra, raster calculator, and Boolean operations, which provide users with a variety of means to combine multiple datasets. In addition, the database structure of GIS allows users to add more descriptive information

Figure 8.3 VR-based Eye-Tracking keeps track of a user's eye movement within immersive space, in this case of a Pompeian house. Through dedicated software, fixation heatmaps are generated interactively based on eye movement and displayed on the 3D model surfaces (Campanaro and Landeschi forthcoming).

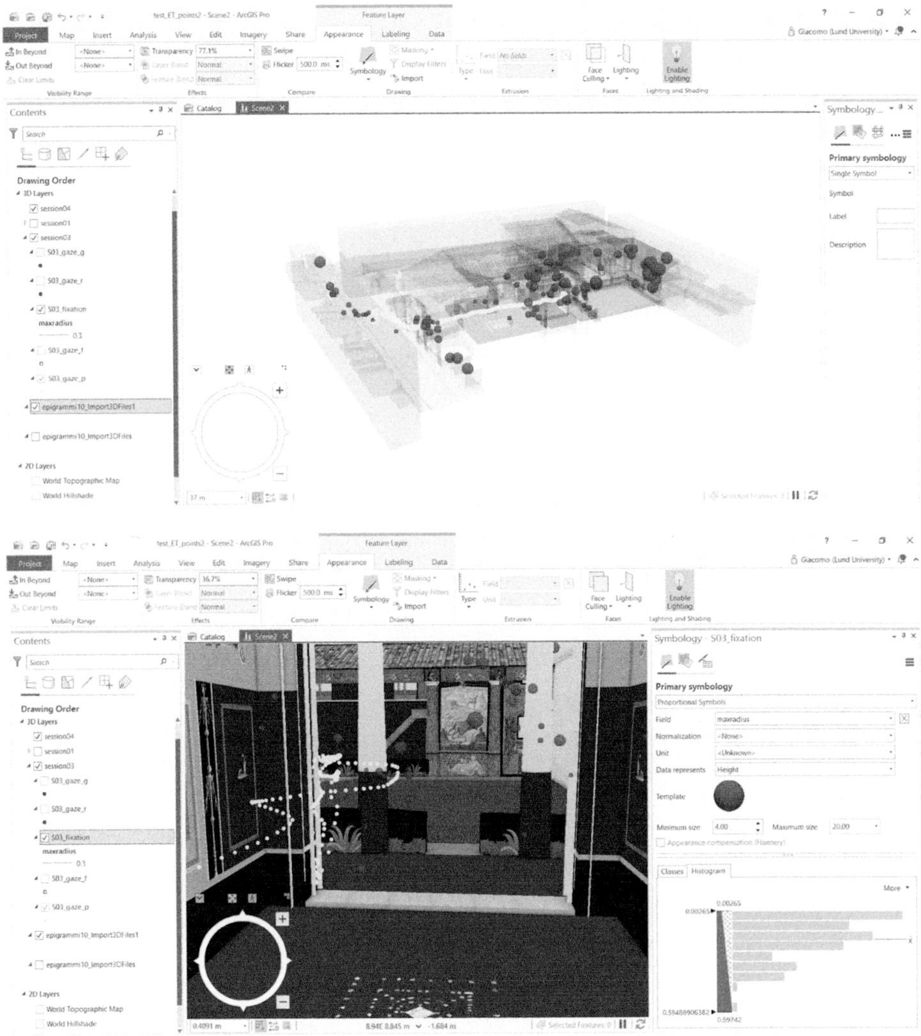

Figure 8.4 ET-derived data are imported in the right spatial location within the model of the Pompeian house, displayed as 3D point shapefiles. In the image above, a movement path marked by yellow points indicates the user's trajectory within the immersive space of the house, along with their fixation points on the different architectural elements of the house, marked by red points (3D model design courtesy of Danilo Campanaro).

that is in itself non-spatial. The user's own description of the visual characteristics of the VR environment is indeed an important support for the 3D visualisation of the ET mapped points. It is therefore crucial to connect spatial information with attribute table entries where a single point of visual attention is accompanied by a more qualitative description of the saliency and the acuity of the observed object (Figure 8.4).

8.2.3 Movement and perception

As noted in Chapter 1, GIS-based analyses of movement began in the 1990s, with the earliest location models aiming at studying paths in prehistoric landscapes. Soon the potential of extending this approach to very diverse research contexts became clear, as it provided a new way to analyse the significance of past landscapes (Gaffney and Stancic 1991; Llobera et al. 2011; Herzog 2014). Similarly to what has been previously described regarding visibility studies, path location models had and still have important limitations due to the very reductionist nature of the simulation scenario, which is digitally created and struggles to properly account for the complexity of the landscape and the complexity of modelling human movement.

Motion Capture technology represents an interesting analytic method that has recently been introduced in the field of experimental archaeology, and is defined as "*the process of recording a live motion event and translating it into usable mathematical terms (...) to obtain a single three-dimensional representation of the performance*" (Menache 2000:1). Although this technique has been widely employed in computer animation and in the game industry, it has not yet been employed that much in archaeological simulations. As Dunn et al. demonstrate (2011), Motion Capture can help archaeologists simulate human movement digitally within a virtual space (representing, for example, an Iron Age hut) where actions connected to daily domestic life are performed by users and analysed in a controlled environment. This allows archaeologists to formulate hypotheses about possible uses of domestic space through the examination of movement data. By conducting this experiment both in the physical space of the reconstructed hut and in its virtually reconstructed counterpart, it is possible to compare different movement paths based on the changing conditions of the task being performed.

Interestingly, movement data represent sequences of temporal actions that can be displayed as a spatial dataset in which each single point represents a state of a human body in a well-defined temporal interval. By examining these data in a GIS environment, it is possible to analyse temporal sequences of actions in relation to a certain environment and better understand how a certain behaviour is affected by external conditions. Again, this reinforces the importance of including human agents in experiments that aim to study aspects of cognition and relationships between past populations and their landscape. In addition, combining multiple user datasets can allow us to detect patterns that help to better understand a certain spatial configuration.

8.2.4 Aurality and acoustics

Another interesting field of application is the analysis of acoustic data. GIS-based analysis of acoustic data has been explored a few times (Paliou and Knight 2013; Primeau and Witt 2017; Cooper 2019; Goodwin and Richards-Rissetto 2020), with experiments involving the combination of integrated analytical approaches where sight and hearing were explored to generate new insights into the social significance of ancient space. As these works demonstrate, archaeoacoustics play an important role in the study of very diverse research contexts, ranging from medieval sacred spaces to Mayan cities.

Sound analysis is usually based on tools that take into account parameters such as terrain elevation values, environmental information (air pressure, temperature) and catchment distances or "soundsheds" (Witt and Primeau 2019) to better understand what the distance within which a certain sound could have been perceived was. However, performing such an analysis completely in GIS has significant limitations connected to the quality of the reconstruction of the original scenario and the number of parameters providing qualitative information about sound propagation.

Additional tools, including 3D modelling software, Virtual Reality and Game Engines now provide users with more sophisticated solutions for coping with the complexity of soundshed analysis. Virtual Reality can be particularly supportive for simulating the human factor. As with ET, it is possible to run experiments in immersive environments where the user's capability to hear different sound sources while moving or standing still in the reconstructed 3D scene is tested. Results from such experiments could be exported and included in a geodatabase, becoming an additional source of information about the spatial configuration of an ancient space.

8.3 Summary and conclusions

To sum up, 3D GIS can now help archaeologists integrate multiple, very diverse datasets, no matter which data source is examined, in the same geospatial location. New sensors and acquisition methods provide unprecedented opportunities for recording novel multi-resolution information for exploring past landscapes in continuity, and even for examining non-material aspects of past realities. Archaeologists can now rely on more robust pipelines for the acquisition, visualisation, and analysis of 3D data in georeferenced space, and such an advance constitutes the foundation for a more reflexive approach and heuristic use of these data in archaeological practice. As the previous chapters have shown, stratigraphic excavations, ancient built spaces, prehistoric caves, and endangered sites are just a few examples of contexts in which a heuristic approach towards the use of 3D geospatial data can increase our understanding of the archaeological record, providing an important asset for archaeological interpretation. It is about time we engage with new challenges in which multiple specialists are involved in archaeological projects and novel pipelines are designed and developed based on specific research questions. Artificial Intelligence, multi-resolution active and passive sensors, and innovative tracking devices constitute only a few examples of research areas that can generate new types of data and new protocols. GIS, in its 3rd and 4th dimension, thus remains a crucial "place to think", a place to relate spatial and non-spatial information so as to build narratives through the examination of multiple, integrated sources.

References

Adkins, L. and Adkins, R. 1989. *Archaeological Illustration.* Cambridge: Cambridge University Press.

Agugiaro, G., Remondino, F., Girardi, G., von Schwerin, J., Richards-Rissetto, H., and de Amicis, R. 2011. Queryarch3D: Querying and visualising 3D models of a Maya archaeological site in a Web-based interface. *Geoinformatics FCE CTU,* 6, pp. 10–17.

Aldenderfer, M. and Maschner, H.D. (eds.). 1996. *Anthropology, Space, and Geographic Information Systems.* Oxford: Oxford University Press.

Alfsdotter, C., Papmehl-Dufay, L., and Victor, H. 2018. A moment frozen in time: Evidence of a late fifth-century massacre at Sandby borg. *Antiquity,* 92(362), pp. 421–436. doi:10.15184/aqy.2018.21.

Ambrosiani, B. 1964. *Fornlämningar och bebyggelse: studier i Attundalands och Södertörns förhistoria.* PhD diss., Uppsala University.

Apel, J. and Storå, J. 2017a. The Pioneer settlements of Gotland: A behavioural ecology approach. In Persson, P., Riede, F., Skar, B., Breivik, H.M. and Jonsson, L. (eds.), *The Ecology of Early Settlement in Northern Europe – Conditions for Subsistence and Survival* (Volume 1). Sheffield: Equinox eBooks Publishing. Dec 2017, pp. 277–309.

Apel, J. and Storå, J. 2017b. Ett återbesök i Stora Förvar och en ny bild av mesolitikum på Gotland. In Wallin, P. and Martinsson-Wallin, H. (eds.), *Arkeologi på Gotland 2. Tillbakablickar och nya forskningsrön.* Institutionen för arkeologi och antik historia, Uppsala Universitet & Gotlands Museum. Uppsala och Visby, pp. 9–21.

Applbaum, N. and Applbaum, Y.H. 2002. The use of medical Computed Tomography (CT) in the study of the ceramic technology of figurines. In Garfinkel, Y. and Miller, M.A. (eds.), *Sha'ar Hagolan 1: Neolithic art in context.* Oxford: Oxbow, pp. 214–220.

Applbaum, N. and Applbaum, Y.H. 2005. The use of medical computed tomography (ct) imaging in the study of ceramic and clay artefacts from the ancient near east. In Uda M., Demortier G., Nakai I. (eds), *X-Rays and Archaeology.* Dordrecht: Springer. https://doi.org/10.1007/1-4020-3581-0_17

Auer, M., Agugiaro, G., Billen, N., Loos, L. and Zipf, A. 2014. Web-based Visualization and Query of semantically segmented multiresolution 3D Models in the Field of Cultural Heritage. In: *ISPRS Annals of the Photogrammetry, Remote Sensing and Spatial Information Sciences,* II-5, doi:10.5194/isprsannals-II-5-33-2014, pp. 33–39.

Barceló, J.A. and Pallarés, M. 1996. A critique of G.I.S. in archaeology. From visual seduction to spatial analysis. *Archeologia e Calcolatori,* 7, pp. 313–326.

Barceló, J.A. 2000. Visualizing what might be. An introduction to virtual reality in archaeology. In Barceló, J.A., Forte, M., and Sanders (eds.), *Virtual Reality in Archaeology.* Oxford (GB): British Archaeological Reports, pp. 9–36.

Barceló, J.A., De Castro, O., Traver, D., and Vicente, O. 2003. A 3D model of an archaeological excavation. In Doerr, M. and Sarris, A. (eds.), *The Digital Heritage of Archaeology. Computer Applications and Quantitative Methods in Archaeology* (1 ed.), Archive of Monuments and Publications, pp. 85–87.

Barceló, J.A. and Vicente, O. 2004. Some problems in archaeological excavation 3D modelling. In Ausserer, K.F., Borner, W., Goriany, M., and Karlhuber-Vockl, L. (eds.), *Enter the Past. The e-way into the Four Dimensions of Cultural Heritage. CAA 2003*. BAR International Series, vol. 1227. Oxford: Archaeopress, pp. 400–404.

Barceló, J. 2014. 3D modelling and shape analysis in archaeology. In Remondino, F. and Campana, S. (eds.), *3D Recording and Modelling in Archaeology and Cultural Heritage Theory and Best Practice*. Oxford: BAR publishing, 15–22.

Barker, P. 2002. *Techniques of Archaeological Excavation*, 3rd edition. London & New York: Routledge.

Barry, A.M. 1997. *Visual Intelligence: Perception, Image and Manipulation in Visual Communication*. Albany: Suny Press.

Benefiel, R.R. 2010. Dialogues of ancient graffiti in the House of Maius Castricius in Pompeii. *American journal of archaeology*, 114(1), pp. 59–101.

Berggren, Å, Dell'Unto, N., Forte, M., Haddow, S., Hodder, I., Issavi, J., Lercari, N., Mazzucato, C. Mickel, A., and Taylor, J. 2015. Revisiting reflexive archaeology at Çatalhöyük: Integrating digital and 3D technologies at the trowel's edge. *Antiquity*, 89(344), pp. 433–448. doi:10.15184/aqy.2014.43

Bernardini, W., Barnash, A., Kumler, M., and Wong, M. 2013. Quantifying visual prominence in social landscapes. *Journal of Archaeological Science*, 40, pp. 3946–3954.

Bewley, R. 2005. Aerial archaeology. The first century. In Bourgeois J. and Meganck M. (eds.), *Aerial Photography and Archaeology 2003. A Century of Information*. Ghent: Ghent University Press, 15–30.

Bewley, R., Wilson, A.I., Kennedy, D., Mattingly, D., Banks, R., Bishop, M., Bradbury, J., Cunliffe, E., Fradley, M., Jennings, R., Mason, R., Rayne, L., Sterry, M., Sheldrick, N., and Zerbini, A. 2016. Endangered archaeology in the Middle East and North Africa: Introducing the EAMENA Project, In Campana, S., Scopigno, R., Carpentiero, G., and Cirillo, M. (eds.), *CAA2015. Keep the Revolution Going: Proceedings of the 43rd Annual Conference on Computer Applications and Quantitative Methods in Archaeology*. Oxford: Archaeopress Archaeology, pp. 919–932.

Bezzi, A., Bezzi, L., Francisci, D., and Gietl, R. 2006. L'utilizzo di voxel in campo archeologico, *Geomatic workbooks 6*. Resource document http://www.arc-team.com/Upload/Geom_Workbooks6_voxel_archeologia_it.pdf. Accessed 23 February 2021.

Bezzi, L. 2012. 4D GIS. In *ATOR (Arc-Team Open Research) blog*, Friday, 16 March 2012. Resource document https://arc-team-open-research.blogspot.com/search?q=gvSIG. Accessed 23 February 2021.

Binford, S.R. and Binford, L.R. 1968. *New Perspectives in Archeology*. Chicago: Aldine Pub. Co.

Braidotti, R. 2016. Posthuman critical theory. In Banerji, D., and Makarand, R.P., (eds.), *Critical Posthumanism and Planetary Futures*. New Delhi: Springer, pp. 13–32.

Brandt, R., Groenwoudt, B.J., and Kvamme, K.L. 1992. An experiment in archaeological site location: Modeling in the Netherlands using GIS techniques. *World Archaeology*, Analytical Field Survey, 24(2), pp. 268–282.

Brinch Petersen, E. 2015. Diversity of mesolithic Vedbaek. *Acta Archaeologica*, 86(1), pp. 19–202.

Buccellati, F. 2016. *Three-Dimensional Volumetric Analysis in an Archaeological Context: The Palace of Tupkish at Urkesh and Its Representation*. Malibu: Undena.

Buchner, E. 1976. *Solarium Augusti und Ara Pacis*, Mainz am Rhein: Philipp von Zabern 83, pp. 319–365.

Buckland, P., Dell'Unto, N., and Palsson, G. 2018. To tree, or not to tree? On the empirical basis for having past landscapes to experience. *Digital Humanities Quarterly*, 12(2). http://digitalhumanities.org/dhq/vol/12/3/000383/000383.html.

Callieri, M., Dell'Unto, N., Dellepiane, M., Scopigno, R., Söderberg, B., and Larsson, L. 2011. Documentation and interpretation of an archeological excavation: An experience with dense stereo reconstruction tools. In Dellepiane, M., Nicolucci, F., Pena Serna, Rushmeier, H., and Van Gool, L. (eds.), *VAST2011 International Symposium on Virtual Reality, Archaeology and Cultural Heritage*. Fraunhofer: Eurographics – European Association for Computer Graphics, pp. 33–40.

Campana, S. 2014. 3D modelling in archaeology and cultural heritage-theory and best practice. In Campana, S. and Remondino, F. (eds.), *3D Recording and Modelling in Archaeology and Cultural Heritage. Theory and Best Practices*. Oxford: Archaeopress, pp. 113–127. (BAR International Series 2598).

Campana, S., 2016. *Archaeology, Remote Sensing. Encyclopedia of Geoarchaeology*. Dordrecht/Heidelberg/New York/London: Springer, pp. 703–724.

Campana, S. 2017. Emptyscapes: Filling an 'Empty' Mediterranean Landscape at Rusellae, Italy. *Antiquity*, 91(359), pp. 1223–1240.

Campana, S. and Piro, S. eds., 2008. *Seeing the Unseen. Geophysics and Landscape Archaeology*. London: CRC Press. Taylor & Francis Group.

Campanaro, D.M., Landeschi, G., Dell'Unto, N., and Leander Touati, A-M. 2015. 3D GIS for cultural heritage restoration: A 'White Box' Workflow. *Journal of Cultural Heritage*, 18, pp. 321–332. https://doi.org/10.1016/j.culher.2015.09.006

Campanaro, D.M. and Landeschi, G. (in press). Re-viewing Pompeian domestic space through combined Virtual Reality-based Eye Tracking and 3D GIS. *Antiquity*.

Campiani, A., Lingle, A., and Lercari, N. 2019. Spatial analysis and heritage conservation: Leveraging 3D data and GIS for monitoring earthen architecture. *Journal of Cultural Heritage*. doi:10.1016/j.culher.2019.02.011.

Carpentier, F. and Lundin J. 2015. Building 6,24 and 17 (Çatalhöyük 2015) Archive Report, *Report by the Catalhoyuk Research Project 44–8*. Retrieved from Çatalhöyük Research Project. www.catalhoyuk.com/research/archive_reports

Castleford, J. 1991. Archaeology, GIS, and the time dimension: An overview. In: Lock, G. and J. Moffett (eds.), CAA91. *Computer Applications and Quantitative Methods in Archaeology* (BAR International Series S577). Oxford: Tempus Reparatum, pp. 95–106.

Chapman, H., 2006. *Landscape archaeology and GIS*. Tempus.

Church, T. J. Brandon, G. Burgett 1999. GIS Applications in Archaeology: Method in Search of Theory, in Wescott, K. and Brandon, R. (eds.), *Practical Applications of GIS for Archaeologists*: A Predictive Modelling Kit. Taylor and Francis, London.

Cignoni, P., Callieri, M., Corsini, M., Dellepiane, M., Ganovelli, F., and Ranzuglia, G. 2008.

Clarke, D.L. 1968. *Analytical Archaeology*. London: Methuen.

Clarke, D.L. and Clarke, D. (eds.) 1977. *Spatial Archaeology*. Cambridge: Academic Press.

Conolly, J. and Lake, M. 2006. *Geographical Information Systems in Archaeology*. Cambridge: Cambridge University Press.

Continenza, R., Trizio, I., Giannangeli, A., and Tata, A. 2016. HBIM for restoration projects: case-study on San Cipriano Church in Castelvecchio Calvisio, Province of L'Aquila, Italy. *Disegnarecon*, 9(16), pp. 315–323.

Cooper, C. 2019. The sound of debate in Georgian England: Auralising the house of commons. *Parliamentary History*, 38(1), pp. 60–73.

Cosmacini, P. and Piacentini, P. 2008. Notes on the history of the radiological study of Egyptian mummies: From X-rays to new imaging techniques. *La Radiologia Medica*, 113(5), pp. 615–626. doi:10.1007/s11547-008-0280-7

de Cambray, B. 1993. Three-dimensional (3D) modelling in a geographical database. In *Proceedings of the 11th International Symposium on Computer-Assisted Cartography*, Auto-Carto'11, Minneapolis, pp. 338–347.

de Cambray, B. and Yeh, T.S. 1994. A multidimensional (2D, 2,5D, and 3D) geographical data model. In *Proceedings of the Sixth International Conference on Management of Data*, COMAD'94, Bangalore, India.

De Reu, J., Plets, G., Verhoeven, G., Bats, M., Cherretté, B., De Maeyer, W., De Smedt, P., Deconynck, J., Herremans, D., Laloo, P., Van Meirvenne, M., and De Clercq, W. 2013.

Towards a three-dimensional cost-effective registration of the archaeological heritage. *Journal of Archaeological Science*, 40(2), pp. 1108–1121.

De Reu, J., De Smedt, P., Herremans, D., Van Meirvenne, M., Laloo, P., and De Clercq, W. 2014. On introducing an image-based 3D reconstruction method in archaeological excavation practice. *Journal of Archaeological Science*, 41, pp. 251–262.

Dellepiane, M., Dell'Unto, N., Callieri, M., Lindgren, S., and Scopigno, R. 2013. Archeological excavation monitoring using dense stereo matching techniques. *Journal of Cultural Heritage*, 14, pp. 201–210. Elsevier.

Dell'Unto, N., Ferdani, D., Leander Touati, A., Dellepiane, M., Callieri, M., and Lindgren, S. 2013a. Digital reconstruction and visualization in archaeology: Case-study drawn from the work of the Swedish Pompeii Project. In Addison, A., Guidi, G., De Luca, L., and Pescarin, S. (eds.), *Digital Heritage International Congress*, Marseilles, FR: IEEE - Institute of Electrical and Electronics Engineers Inc., pp. 621–628.

Dell'Unto, N., Callieri, M., Dellepiane, M., Leander Touati, A., Lindgren, S., and Larsson, C. 2013b. Pompei revived: Scanning mission- insula V 1. In Earl, G., Sly, T., Chrysanthi, A., Murrieta-Flores, P., Papadopoulos, C., Romanowska, I., and Wheatley, D. (eds.), *The Computer Applications and Quantitative Methods in Archaeology (CAA)*. 26–29 March 2012. Southampton: Amsterdam University Press, 199–207.

Dell'Unto, N. 2014. The use of 3D models for intra-site investigation in archaeology. In Campana, S. and Remondino, F. (eds.), *3D Surveying and Modeling in Archaeology and Cultural Heritage Theory and Best Practices*. Oxford: BAR International Series, pp. 151–158.

Dell'Unto, N., Landeschi, G., Leander Touati, A.-M., Dellepiane, M., Callieri, M., and Ferdani, D. 2015. Experiencing ancient buildings from a 3D GIS perspective: A case drawn from the Swedish Pompeii Project. *Journal of Archaeological Method and Theory*, 23, pp. 73–94. Plenum Press.

Dell'Unto, N. 2016. Using 3D GIS platforms to analyse and interpret the past. In Forte, M. and Campana, S. (eds.), *Digital Methods and Remote Sensing in Archaeology: Archaeology in the Age of Sensing (Quantitative Methods in the Humanities and Social Sciences)*. Springer, pp. 305–322. doi:10.1007/978-3-319-40658-9_14

Dell'Unto, N., Landeschi, G., Apel, J., and Poggi, G. 2017. 4D recording at the Trowel's edge: Using three-dimensional simulation platforms to support field interpretation. *Journal of Archaeological Science: Reports*, 12, pp. 632–645. doi:10.1016/j.jasrep.2017.03.011

Dell'Unto, N. 2018. 3D models and knowledge production. In Huvila, I. (ed.), *Archaeology and Archaeological Information in the Digital Society*. Oxon and New York: Routledge, pp. 54–69.

Dell'Unto, N. 2020. The analytical role of 3D realistic computer graphics. In Gillings, M., Hacigüzeller, P., and Lock, G. (eds.), *Archaeological Spatial Analysis: A Methodological Guide*. Oxon and New York Routledge, pp. 444–459.

Demetrescu, E. 2015. Archaeological stratigraphy as a formal language for virtual reconstruction. Theory and practice. *Journal of Archaeological Science*, 57, pp. 42–55.

Demetrescu, E. and Ferdani, D. 2021. From field archaeology to virtual reconstruction: A five steps method using the extended matrix. *Applied Science*, 11, p. 5206. doi:10.3390/app11115206

Der Manuelian, P. 2013. Giza 3D: Digital archaeology and scholarly access to the Giza Pyramids: The Giza Project at Harvard University. In *Proceedings of Digital Heritage 2013* (Digital Heritage International Congress), Marseille, France, 28 October–1 November 2013, vol. 2. Institute of Electrical and Electronics Engineers, pp. 727–734.

Derudas, P., Dell'Unto, N., Callieri, M., and Apel, J. 2021. Sharing archaeological knowledge: The Interactive Reporting System. *Journal of Field Archaeology*, 46(5), 303–315. https://doi.org/10.1080/00934690.2021.1911132

Derudas, P. and Berggren, Å. 2021. Expanding field-archaeology education: The integration of 3d technology into archaeological training. *Open Archaeology*, 7(1), 556–573. https://doi.org/10.1515/opar-2020-0146

Díaz-Guardamino, M. and Morgan, C. 2019. Human, transhuman, posthuman digital archaeologies: An introduction. *European Journal of Archaeology*, 22(3), pp. 320–323.

Dodd, J. 2018. The application of high-performance computing in rock art documentation and research. *Adoranten*, n 1, pp. 92–104.

Doneus, M. and Neubauer, W. 2005. 3D laser scanners on archaeological excavations. In Dequal, S. (ed.), *Proceedings of the XXth International Symposium CIPA*, Torino 2005. The International Archives of Photogrammetry, Remote Sensing and Spatial Information Sciences, XXXVI-5/C34/1, pp. 226–231.

Drebin, R.A., Carpenter, L., and Hanrahan, P. 1988. Volume rendering. *ACM Siggraph Computer Graphics*, 22(4), pp. 65–74.

Ducke, B., Score, D., and Reeves, J. 2011. Multiview 3D reconstruction of the archaeological site at Weymouth from image series. *Computers & Graphics*, 35, pp. 375–382. doi:10.1016/j.cag.2011.01.006.

Dunn, S., Woolford, K., Barker, L., Norman, S.J., White, M., Hedges, M., Bailey, H., Fulford, M., and Clarke, A. 2011. Motion in place: A case study of archaeological reconstruction using motion capture. In Zhou, M., I. Romanowska, Z. Wu, P. Xu and P. Verhagen (eds.), *Revive the Past. Computer Applications and Quantitative Methods in Archaeology (CAA)*. Proceedings of the 39th International Conference, Beijing, April 12-16, Amsterdam: Pallas Publications pp. 98–106.

Earl, G.P. 2007. De-construction sites: Romans and the digital playground. In Bowen, J.P., Keene, S., and MacDonald, L. (eds.), *EVA Conferences International: Conference Proceedings*. Hampton Hill: Middlesex, pp. 1–11.

Earl, G., Martinez, K., and Malzbender, T. 2010. Archaeological applications of polynomial texture mapping: Analysis, conservation and representation. *Journal of Archaeological Science*, 37(8), pp. 2040–2050.

Earl, G., Porcelli, V., Papadopoulos, C., Beale, G., Harrison, M., Pagi, H., and Keay, S. 2013. Formal and informal analysis of rendered space: The basilica Portuense. In Bevan, A. and Lake, M. (eds.), *Computational Approaches to Archaeological Spaces*. Walnut Creek: Left Coast Press, pp. 265–305.

Edgeworth, M. 2014. From spade-work to screen-work: New forms of archaeological discovery in digital space. In Carusi, A., Hoel, A., Webmoor, T., and Woolgar, S. (eds.), *Visualization in the Age of Computerization*. New York and Oxon: Routledge, pp. 40–58.

Eiteljorg, II, H., Fernie, K., Huggett, J., and Robinson, D. 2002. CAD: A guide to good practice. *AHDS Guides to Good Practice*. http://adsahds.ac.uk/project/goodguides/cad.

ESRI. 2012. The multipatch geometry type – An ESRI white paper December 2008. Resource document. http://www.esri.com/library/whitepapers/pdfs/multipatch-geometry-type.pdf. Accessed 26 April 2016.

ESRI. 2015. GIS Tutorial 10- ArcGIS 3D Analyst. Resource document. http://downloads2.esri.com/ESRIpress/images/144/GISTutorial_samplechapter.pdf. Accessed 22 June 2021.

ESRI. 2016. ArcGIS 3D Analyst. Resource document. http://www.esri.com/software/arcgis/extensions/3danalyst. Accessed 22 June 2016.

ESRI. 2018. https://desktop.arcgis.com/en/arcmap/10.4/extensions/3d-analyst/what-is-a-functional-surface-.htm#GUID-75B7B08A-544D-4F31-B2F5-C958FB9E422A. Accessed 22 June 2021.

ESRI. 2020. (SLOPE). https://desktop.arcgis.com/en/arcmap/10.3/tools/spatial-analyst-toolbox/slope.htm#:~:text=Slope%20is%20the%20rate%20of,the%20ground%20x%2Cy%20 units. Accessed 22 May 2021.

ESRI 2020. What is Empirical Bayesian Kriging 3D? https://pro.arcgis.com/en/pro-app/latest/help/analysis/geostatistical-analyst/what-is-empirical-bayesian-kriging-3d-.htm. Accessed 22 June 2021.

Fahlander, F. and Kjellström, A. 2010. Beyond sight: Archaeologies of sensory perception. In Fahlander, F. and Kjellström, A. (eds.), *Making Sense of Things: Archaeologies of Sensory Perception*. Department of Archaeological and Classical History, Stockholm: Stockholm University, pp. 1–13.

Ferdani, D., Demetrescu, E., Cavalieri, M., Pace, G., and Lenzi, S. 2019. 3D modelling and visualization in field archaeology. From survey to interpretation of the past using digital technologies. *Groma* 4. doi:10.12977/groma26.

Ferdani, D., Fanini, B., Piccioli, M.C., Carboni, F., and Vigliarolo, P. 2020. 3D reconstruction and validation of historical background for immersive VR applications and games: The case study of the Forum of Augustus in Rome. *Journal of Cultural Heritage*. Elsevier. In press.

Fleming, A. 2006. Post-processual landscape archaeology: A critique. *Cambridge Archaeological Journal*, 16(3), p. 267.

Foley, J.D., van Dam, A., Feiner, S.K., and Hughes, J.F. (eds.). 1990. *Computer Graphics: Principles and Practice*, 2nd edition. Boston: Addison-Wesley Publishing Company.

Foley, J.D. 1996. 12.7 Constructive solid geometry. In Foley, J.D., van Dam, A., Feiner, S.K., and Hughes, J. (eds.), *Computer Graphics: Principles and Practice*. Boston: Addison-Wesley Professional, pp. 557–558. ISBN 9780201848403.

Forte, M. and Silotti, A. (eds.) 1997. *Virtual Archaeology. Re-Creating Ancient Worlds*. New York: Abrams, 288 p.

Forte, M. 2008. Virtual archaeology: Communication in 3D and ecological thinking. In Frischer, B. and Dakouri-Hild, A. (eds.), *Beyond Illustration: 2D and 3D Digital Technologies as Tools for Discovery in Archaeology*. Oxford: Archaeopress, pp. 20–34. (British Archaeological Reports S1805).

Forte, M. (ed.) 2010. *Cyber-archaeology*. BAR International Series 2177. Oxford: Archaeopress.

Forte, M. 2011. Cyber-archaeology: Notes on the simulation of the past. *Virtual Archaeology Review*, 2(4), pp. 7–18.

Forte, M., Dell'Unto, N., Issavi, J., Onsurez, L., and Lercari, N. 2012. 3D archaeology at Çatalhöyük. *International Journal of Heritage in the Digital Era*, 1(3), pp. 351–377. doi:10.1260/2047-4970.1.3.351

Forte, M. 2014. Virtual reality, cyberarchaeology, teleimmersive archaeology. In Campana, S. and Remondino, F. (eds.), *3D Surveying and Modeling in Archaeology and Cultural Heritage. Theory and Best Practices*. Oxford: BAR Archaeopress, pp. 115–129.

Forte, M., Dell'Unto, N., Jonsson, K., and Lercari, N. 2015. Interpretation process at Çatalhöyük using 3D. In Hodder, I. and Marciniak, A. (eds.), *Assembling Çatalhöyük, 1. European Association of Archaeologists Contemporary Themes in Archaeology*, Leeds: Maney Publishing, pp. 43–57.

Fredengren, C. 2015. Nature: Cultures: Heritage, sustainability and feminist posthumanism. *Current Swedish Archaeology*, 23, pp. 109–130.

Frieman, C. and Gillings, M. 2007. Seeing is perceiving? *World Archaeo*logy, 39(1), pp. 4–16.

Frischer, B. 2008. From digital illustration to digital heuristics. In Frischer, B. and Dakouri-Hild, A. (eds.), *Beyond Illustration: 2D and 3D Digital Technologies as Tools for Discovery in Archaeology*. Oxford: Archaeopress, pp. v–xxii. (British Archaeological Reports S1805).

Frischer, B. and Fillwalk, J. 2013. A computer simulation to test the Buchner thesis: The relationship of the Ara Pacis and the meridian in the Campus Martius, Rome. In *2013 Digital Heritage International Congress (DigitalHeritage)*, Vol. 1. IEEE, pp. 341–345.

Gaffney, V. and Stancic, Z. 1991. GIS approaches to regional analysis: A case study of the Island of Hvar. *American Journal of Archaeology*, 97(1993), p. 799.

Gaffney, V.L., Fitch, S. and Smith, D.N., 2009. Europe's lost world: the rediscovery of Doggerland. Council for British Archaeology Research Report (Vol. 160). Council for British Archaeology St Mary's House, 66 Bootham, York.

Galeazzi, F. and Richards-Rissetto, H. 2018. Editorial introduction: Web-based archaeology and collaborative research. *Journal of Field Archaeology*, 43(sup1), pp. S1–S8, doi:10.1080/00934690.2018.1512701

Garagnani, S., Gaucci, A., and Gruška, B. 2016. From the archaeological record to archaeobim: The case study of the Etruscan temple of Uni in Marzabotto. *Virtual Archaeology Review*, 7, 77. doi:10.4995/var.2016.5846.

Gavryushkina, M. 2018. Layer by (3D) layer, 3D GIS stratigraphic analysis of Chlorakas-Palloures, Cyprus. MA diss. thesis. University of Leiden.

Gavryushkina, M. 2021. The potential and problems of volumetric 3D modeling in archaeological stratigraphic analysis: A case study from Chlorakas-Palloures, Cyprus. *Digital Applications in Archaeology and Cultural Heritage*, 21, p.e00184.

Gay, E., Cooper, D., Kimia, B., Taubin, G., Cabrini, D., Karumuri, S., Doutre, W., Liu, S., Galor, K., Sanders, D., and Willis, A. 2010. REVEAL intermediate report. In *2010 IEEE Computer Society Conference on Computer Vision and Pattern Recognition-Workshops*. IEEE, pp. 1–6.

Gillings, M. and Goodrick, G.T. 1996. Sensuous and reflexive GIS: Exploring visualization and VRML. *Internet Archaeology*, 1. Resource document. http://intarch.ac.uk/journal/issue1/gillings_index.html. Accessed 20 March 2014.

Gillings, M. and Wheatley, D. 2002. *Spatial Technology and Archaeology: The Archaeological Applications of GIS*. London and New York: Routledge.

Gillings, M. 2012. Landscape phenomenology, GIS and the role of affordance. *Journal of Archaeological Method and Theory*, 19(4), pp. 601–611.

Gillings, M. 2015a. Betylmania?–Small standing stones and the Megaliths of South-West Britain. *Oxford Journal of Archaeology*, 34(3), pp. 207–233.

Gillings, M. 2015b. Mapping invisibility: GIS approaches to the analysis of hiding and seclusion. *Journal of Archaeological Science*, 62, pp. 1–14.

Gillings, M. 2017. Mapping liminality: Critical frameworks for the GIS-based modelling of visibility. *Journal of Archaeological Science*, 84, pp. 121–128.

Gillings, M., Hacıgüzeller, P. and Lock, G., 2020. *Archaeology and Spatial Analysis. In Archaeological Spatial Analysis*. Oxon: Routledge, pp. 1–16.

Giuliani, C.F. 2008. Prefazione. In Bianchini, M. (ed.), *Manuale di rilievo e di documentazione digitale in archeologia* (Vol. 362). Roma: Aracne editrice, pp. 9–12.

Goodchild, M.F. 1991. Geographic information systems. *Progress in Human Geography*, 15(2), pp. 194–200.

Goodman, D. and Piro, S., 2013. *GPR Remote Sensing in Archaeology* (Vol. 9, p. 233). New York: Springer.

Goodwin, G. and Richards-Rissetto, H. 2020. Modelling acoustics in ancient maya cities: Moving towards a synesthetic experience using GIS & 3D simulation. In *CAA: Digital Archaeologies, Material Worlds (Past and Present)*. doi:10.15496/publikation-43230.

Gowlett, J.A.J. 1997. High-definition archaeology: Ideas and evaluation. *World Archaeology*, 29(2), pp. 152–171.

Graceprojektet. 2014. Resource document. http://www.graceprojektet.se/en/. Accessed 22 June 2016.

Grilli, E. and Remondino, F. 2019. Classification of 3D digital heritage. *Journal of Remote Sensing*, 11. doi:10.3390/rs11070847.

Grøn, O., Palmér, S., Stylegar, F.A., Esbensen, K., Kucheryavski, S., and Aase, S. 2011. Interpretation of archaeological small-scale features in spectral images. *Journal of Archaeological Science*, 38(9), pp. 2024–2030. doi:10.1016/j.jas.2009.11.023.

Hacıgüzeller, P. 2012. GIS, critique, representation and beyond. *Journal of Social Archaeology*, 12(2), pp. 245–263.

Haggett, P. 1977. Geography in a steady-state environment. *Geography*, 62(3), pp. 159–167.

Hamilakis, Y. 2014. *Archaeology and the Senses: Human Experience, Memory, and Affect*. Cambridge: Cambridge University Press.

Harris, E. 1989. *Principles of Archaeological Stratigraphy*. 2nd edition. London: Academic Press. doi:10.1007/s007690000247

Harris, T.M. and Lock, G.R. 1990. The diffusion of a new technology: A perspective on the adoption of geographic information systems within UK archaeology. In Allen, K.M.S., Green, S.W., Zubrow, E.B.W. (eds.), *Interpreting Space: GIS and Archaeology*. London: Taylor & Francis, pp. 33–54.

Harris, T.M. and Lock, G.R. 1996. Multi-dimensional GIS: Exploratory approaches to spatial and temporal relationships within archaeological stratigraphy. In Kamermans, H. and

Fennema, K. (eds.), *Interfacing the Past. Computer Applications and Quantitative Methods in Archaeology CAA95*. Vol. II (Analecta Praehistorica Leidensia 28). Leiden: Institute of Prehistory, University of Leiden, pp. 307–316.

Henriksson, M. and Nilsson, B. 2015. Hästholmen 1:67 Torhamns socken, Karlskrona kommun, Arkeologisk efterundersökning. *Blekinge Museum Rapport 2015:4*. Karlskrona.

Hermon, S. 2008. Reasoning in 3D: A critical appraisal of the role of 3D modelling and virtual reconstructions in archaeology. In Frischer, B. and Dakouri-Hild, A. (eds.), *Beyond Illustration: 2D and 3D Digital Technologies as Tools for Discovery in Archaeology*. Oxford: Archaeopress, pp. v–xxii. (British Archaeological Reports S1805).

Herzog, I. 2014. Least-cost paths–some methodological issues. *Internet Archaeology*, 36. doi:10.11141/ia.36.5, https://intarch.ac.uk/journal/issue36/herzog_index.html

Hillier, B. and Hanson, J. 1989. *The Social Logic of Space*. Cambridge: Cambridge University Press.

Historic England. 2017. *Photogrammetric Applications for Cultural Heritage. Guidance for Good Practice*. Swindon: Historic England.

Hodder, I. and Orton, C. 1976. *Spatial Analysis in Archaeology*. Cambridge: Cambridge University Press.

Hodder, I. 1997. Always momentary, fluid and flexible: Towards a reflexive excavation methodology. *Antiquity*, 71(273), pp. 691–700.

Hodder, I. (ed.) 2000. *Towards a Reflexive Method in Archaeology: The Example of Çatalhöyük*. Cambridge: McDonald Institute for Archaeological Research/British Institute of Archaeology at Ankara Monograph.

Hodder, I. 2007. Çatalhöyük in the context of the Middle Eastern Neolithic. *Annual Review of Anthropology*, 36, pp. 105–120.

Horn, C., Pitman, D., and Potter, R. 2019. An evaluation of the visualisation and interpretive potential of applying GIS data processing techniques to 3D rock art data. *Journal of Archaeological Science: Reports*, 27, 1–13.

Howorth, J.P. 2020. *Archaeobotany in the 4th Dimension: Visualising Specialist Data in a 3D GIS*. MA diss. thesis. Lund University. Department of Archaeology & Ancient History. Accessed 5 May 2021.

Huggett, J. 2017. The apparatus of digital archaeology. *Internet Archaeology*, 44. http://doi.org/10.11141/ia.44.7

Huggett, J. 2021. Algorithmic agency and autonomy in archaeological practice. *Open Archaeology*, 7(1), pp. 417–434. https://doi.org/10.1515/opar-2020-0136

Hughes, P.J. and Lampert, R.J. 1977. Occupational disturbance and types of archaeological deposit. *Journal of Archaeological Science*, 4(2), pp. 135–140.

Hughes, S. 2011. CT scanning in archaeology, computed tomography – Special applications. In Saba, L. (ed.), *InTech Europe*. Resource document: http://www.intechopen.com/books/computedtomography-special-applications/ct-scanning-in-archaeology.

Hughes, J., Van Dam, A., MaGuire, M., Sklar, D., Foley, J., Feiner, S., and Akeley, K. 2014. *Computer Graphic Principles and Practice*. Boston: Addison-Wesley.

Huisman, D.J., Ngan-Tillard, D., Tensen, M.A., Laarman, F.J., and Raemaekers, D.C.M. 2014. A question of scales: Studying Neolithic subsistence using micro CT scanning of midden deposits. *Journal of Archaeological Science*, 49, p. 585.

Huvila, I. 2018. The subtle difference between knowledge and 3D knowledge. *Hamburger Journal für Kulturanthropologie (HJK)*, (7), pp. 99–111. Abgerufen von https://journals.sub.uni-hamburg.de/hjk/article/view/1196

Ingold, T. 1993. The temporality of the landscape. *World Archaeology*, 25(2), pp. 152–174.

Janz, B. B. (2018). Virtual place and virtualized place. In E. Champion (ed.), The Phenomenology of Real and Virtual Places, 1st edn. New York: Routledge, pp. 60–75. https://doi.org/10.4324/9781315106267-6

Jensen, P. 2017. Where are we? Reviewing the integration of complex spatial data in current field archaeology. *Internet Archaeology*, 44. https://doi.org/10.11141/ia.44.8

Jensen, P. 2018. Semantically enhanced 3D: A web-based platform for spatial integration of excavation documentation at Alken Enge, Denmark. *Journal of Field Archaeology*, 43(sup1), pp. S31–S44. doi:10.1080/00934690.2018.1510299

Johansson, N. and Johansson, L.G. 2001. Rescue archaeology. *Archaeology – Vol II*, EOLSS. Resource document: http://www.eolss.net/SampleChapters/C04/E6-21-04-04.pdf. Accessed 10 July 2013.

Jones, C. B. (1989), Data structures for three-dimensional spatial information systems in geology, *International Journal of Geographical Information Systems*, 3(1), pp. 15–31.

Katsianis, M., Tsipidis, S., Kotsakis, K., and Kousoulakou, A. 2008. A 3D digital workflow for archaeological intra-site research using GIS. *Journal of Archaeological Science*, 35(3), pp. 655–667.

Katsianis, M., Tsipidis, S., and Kalisperakis, I. 2015. Enhancing excavation archives using 3D spatial technologies. In Papadopoulos, C., Paliou, E., Chrysanthi, A., Kotoula, E., and Sarris, A. (eds.), *Archaeological Research in the Digital Age. Proceedings of the 1st Conference on Computer Applications and Quantitative Methods in Archaeology Greek Chapter (CAA-GR)*. Rethymno, 6–8 March 2014. Rethymno: IMS-FORTH, pp. 46–54.

Katsianis, M., Kotsakis, K., and Stefanou, F. 2021. Reconfiguring the 3D excavation archive. Technological shift and data remix in the archaeological project of Paliambela Kolindros, Greece. *Journal of Archaeological Science: Reports*, 36, p. 102857. ISSN 2352-409X. doi:10.1016/j.jasrep.2021.102857.

Kaufman, A., Cohen, D., and Yagel, R. 1993. Volume graphics. *Computer*, 26(7), pp. 51–64.

Kazzazi, S.M. and Kranioti, E.F. 2017. A novel method for sex estimation using 3D computed tomography models of tooth roots: A volumetric analysis. *Archives of Oral Biology*, 83, pp. 202–208.

Klein, M., Vermeulen, F. and Corsi, C., 2012. Radiography of the Past—Three Dimensional, Virtual Reconstruction of a Roman Town in Lusitania. *International Journal of Heritage in the Digital Era*, 1(1_suppl), pp. 131–135.

Kimball, J. 2014. *3D Delineation. A Modernisation of Drawing Methodology for Field Archaeology*. MA diss. thesis. Lund University. Department of Archaeology & Ancient History. Accessed 5 May 2021.

Kintigh, K.W. and Ammerman, A.J., 1982. Heuristic approaches to spatial analysis in archaeology. American Antiquity, 47(1), pp.31-63.

Klimowicz, A. and Tung, B. 2015. Plaster head installation: An overview of the discovery. In *Çatalhöyük 2015 Archive Report*. Resource document: http://www.catalhoyuk.com/sites/default/files/media/pdf/Archive_Report_2015.pdf. Accessed 5 March 2021.

Klinkenberg, V. 2016. Are we there yet? 3D GIS in archaeological research, the case of Tell Sabi Abyad, Syria. In Kamermans, H., de Neef, W., Piccoli, C., Posluschny, A., and Scopigno, R. (eds.), *The Three Dimensions of Archaeology*. Proceedings of the XVII UISPP World Congress no. 7. Oxford: Archaeopress Publishing Ltd. pp. 39–48.

Knüsel, C.J., Haddow, S.D., Sadvari, J.D., Dell'Unto, N., and Forte, M. 2013. Bioarchaeology in 3D: Three-dimensional modeling of human burials at Neolithic Çatalhöyük. Poster presented at *82nd Meeting of the American Association of Physical Anthropologists*, Knoxville. 9–13 April 2013. doi:10.13140/RG.2.2.18447.69287.

Kockel, V. 2004. Towns and tombs. Three dimensional documentation of archaeological sites in the Kingdom of Naples in the late 18th century and early 19th century. In Ilaria Bignamini (ed.), *Archives & Excavations: Essays on the History of Archaeological Excavations in Rome and Southern Italy from the Renaissance to the Nineteenth Century* (Vol. 14). Rome: British School at Rome | Series: Archaeological Monographs of the British School at Rome, pp. 143–162.

Koehl, M. and Lott, C., 2008. De la maquette 3D au SIG 3D pour la gestion du patrimoine archéologique. Géomatique expert, n°64, août-septembre 2008, p. 26–35.

Kohn, C.F. 1970. The 1960's: A decade of progress in geographical research and instruction. *Annals of the Association of American Geographers*, 60(2), pp. 211–219.

Kozatsas, J., Kotsakis, K., Sagris, D., and David, K. 2018. Inside out: Assessing pottery forming techniques with micro-CT scanning. An example from Middle Neo-lithic Thessaly. *Journal of Archaeological Science*, 100, pp. 102–119.

Kvamme, K.L. 1983. Computer processing techniques for regional modeling of archaeological site locations. *Advances in Computer Archaeology*, 1(1), pp. 26–52.

Kvamme, K.L. 1990. The fundamental principles and practice of predictive archaeological modelling. In Voorrips, A. (ed.), *Mathematics and Information Science in Archaeology*, Volume 3. Bonn: Holos Verlag, pp. 257–295.

Kvamme, K.L. 1992. A predictive site location model on the High Plains: An example with an independent test. *Plains Anthropologist*, 37(138), pp. 19–40.

Lake, M., Patricia W., Woodman, E., and Mithen, S.J. 1998. Tailoring GIS software for archaeological applications: An example concerning viewshed analysis. *Journal of Archaeological Science*, 25(1), pp. 27–38.

Lambers, K., Eisenbeiss, H., Sauerbier, M., Kupferschmidt, D., Gaisecker, T., Sotoodeh, S., and Hanusch, T. 2007. Combining photogrammetry and laser scanning for the recording and modelling of the Late Intermediate Period site of Pinchango Alto, Palpa, Peru. *Journal of Archaeological Science*, 34(10), p.1702e1712.

Landeschi, G., Dell'Unto, N., Ferdani, D., Leander Touati, A-M., and Lindgren, S. 2015. Enhanced 3D-GIS: Documenting Insula V 1 in Pompeii. In Giligny, F., Djindjian, F., Costa, L., Moscati, P., and Robert, S. (eds.), *Proceedings of the 42nd Conference on Computer Applications and Quantitative Methods in Archaeology – Paris 22nd to 25th April*. Oxford: Archaeopress, pp. 349–360.

Landeschi, G., Dell'Unto, N., Lundqvist, K., Ferdani, D., Campanaro, D.M., and Leander Touati, A.-M. 2016a. 3D-GIS as a platform for visual analysis: Investigating a Pompeian house. *Journal of Archaeological Science* (Vol. 65). Academic Press, pp. 103–113.

Landeschi, G., Nilsson, B., and Dell'Unto, N. 2016b. Assessing the damage of an archaeological site: New contributions from the combination of image-based 3D modelling techniques and GIS. *Journal of Archaeological Science: Reports*, 10, pp. 431–440.

Landeschi, G. 2019. Rethinking GIS, three-dimensionality and space perception in archaeology. *World Archaeology*, 51(1), pp. 17–32.

Landeschi, G., Apel, J., Lindgren, S. and Dell'Unto, N., 2018. An exploratory use of 3D for investigating a prehistoric stratigraphic sequence. In (edited by M. Matsumoto and E. Uleberg) CAA2016: Oceans of Data: Proceedings of the 44th Conference on Computer Applications and Quantitative Methods in Archaeology (p. 433). Archaeopress Publishing Ltd. Oxford, UK.

Landeschi, G., Apel, J., Lundström, V., Storå, J., Lindgren, S., and Dell'Unto, N. 2019. Re-enacting the sequence: Combined digital methods to study a prehistoric cave. *Archaeological and Anthropological Sciences*, 11(6), pp. 2805–2819. doi:10.1007/s12520-018-0724-5

Landeschi, G., Lindgren, S., Gerding, H., Papadimitriou, A., and Wallensten, J. 2020. Ancient Hermione revealed: The contribution of high-performance computing and digital methods to the analysis of a hidden cityscape. *Archaeological Prospection*, 27(4), pp. 315–328.

Landeschi, G. and Betts, E. (eds.) (in press). *Capturing the Senses: Sensory Archaeologies and Digital Methods* (Quantitative Methods in the Humanities and Social Sciences). Springer Nature Switzerland AG.

Lanjouw, T. 2016. Discussing the obvious or defending the contested: Why are we still discussing the "scientific value" of 3D applications in archaeology? In Kamermans, H., de Neef, W., Piccoli, C., Poluschny, A.G., and Scopigno, R. (eds.), *The Three Dimensions of Archaeology, Proceedings of the XVII UISPP World Congress*, Burgos, Spain, 1–7 September 2014. Oxford: Archaeopress, pp. 1–11.

Lantmäteriet 2021: Resource document. Accessed 22 May 2021. https://www.lantmäteriet.se/sv/Kartor-och-geografisk-information/geodataprodukter/produktlista/ortofoto/

Larralde, S.L. and Chandler, S.M. 1982. *Archaeological Inventory in the Seep Ridge Cultural Study Tract, Uintah County, Northeastern Utah with a Regional Predictive Model for Site Location (No. 11)*. Bureau of Land Management, Utah State Office.

Larsson, L., Trinks, I., Söderberg, B., Gabler, M., Dell'Unto, N., Neubauer, W., and Ahlström, T. 2015. Interdisciplinary archaeological prospection, excavation and 3D documentation exemplified through the investigation of a burial at the Iron Age settlement site of Uppåkra in Sweden. *Archaeological Prospection*, 22(3), pp. 143–156. doi:10.1002/arp.1504

Lasaponara R. and Masini, N. (eds.) 2012. *Satellite Remote Sensing: A New Tool for Archaeology*. Heidelberg: Springer-Verlag.

Laurence, R. 1995. The organization of space in Pompeii. In Cornell, T. and Lomas, K. (eds.), *Urban Society in Roman Italy*. London: UCL Press, pp. 63–78.

Law, M. and Collins, A. 2018. *Getting to Know ArcGIS Desktop*, Redlands, California: Esri Press.

Leander Touati, A.-M. 2010. Water, well-being and social complexity in insula V 1 a Pompeian city block revisited. *Opuscula*, 3, pp. 105–161.

Lercari, N. and Lingle, A. 2015. Çatalhöyük Digital Preservation Project (Çatalhöyük 2015) Archive Report, *Report by the Catalhoyuk Research Project 242–247*. Retrieved from Çatalhöyük Research Project www.catalhoyuk.com/research/archive_reports

Levy, T.E. Smith, N.G., Najjar, M., DeFanti, T.A., Lin, A.Y.-M., and Kuester, F. 2012. *Cyberarchaeology in the holy land: The future of the past*. Washington, D.C.: Biblical Archaeology Society.

Levy, T.E. and Jones, I.W.N. 2017. *Cyber-Archaeology and Grand Narratives: Digital Technology and Deep-Time Perspectives on Culture Change in the Middle East* (1st. ed.). Cham, Switzerland: Springer Publishing Company, Incorporated.

Lieberwirth U. 2008. Voxel-based 3D GIS: Modelling and analysis of archaeological stratigraphy. In Frischer, B. and Dakouri-Hild, A. (eds.), *Beyond Illustration: 2D and 3D Digital Technologies as Tools for Discovery in Archaeology*. Oxford: Archaeopress, pp. v–xxii. (British Archaeological Reports S1805) (pp. 85–95).

Lihammer, A. 2007. Bortom riksbildningen. Människor, Landskap och makt i sydöstra Skandinavien. *Lund Studies in Historical Archaeology* 7. Lund.

Lin, Y. and Seales, W.B. 2005. Opaque document imaging: Building images of inaccessible texts. In *Tenth IEEE International Conference on Computer Vision (ICCV'05)*, vol. 1. Beijing, China: IEEE, pp. 662–669.

Lingle, A., Dell'Unto, N., Der, L., Doyle, S., Killackey, K., Klimowicz, A., Meskell, L., Parkes, P., and Tung, B. 2015. Painted Plaster Head (Çatalhöyük 2015 Archive Report, report by the Çatalhöyük Research Project 44–8). Source document Catalhoyuk Research Project website: http://www.catalhoyuk.com/research/archive_reports. Accessed 5 March 2021.

Llobera, M. 1996. Exploring the topography of mind: GIS, social space and archaeology. *Antiquity*, 70, pp. 612–622. Oxford.

Llobera, M. 2003. Extending GIS-based visual analysis: The concept of visualscapes. *International journal of Geographical Information Science*, 17(1), pp. 25–48.

Llobera, M. 2007. Reconstructing visual landscapes. *World Archaeology*, 39(1), pp. 51–69.

Llobera, M., Fábrega-Álvarez, P., and Parcero-Oubiña, C. 2011. Order in movement: A GIS approach to accessibility. *Journal of Archaeological Science*, 38(4), pp. 843–851.

Llobera, M. 2012. Life on a pixel: challenges in the development of digital methods within an "interpretive" landscape archaeology framework. *Journal of Archaeological Method and Theory*, 19(4), pp. 495–509.

Lock, G. 1995. Archaeological computing, archaeological theory, and moves towards contextualism. In Huggett, J. and N. Ryan (eds.), *CAA94. Computer Applications and Quantitative Methods in Archaeology 1994* (BAR International Series 600). Oxford: Tempus Reparatum, pp. 12–18.

Lock, G. 2003. *Using Computers in Archaeology, towards Virtual Pasts*. Oxon: Routledge.

Lock, G.R. and Molyneaux, B. 2007. *Confronting Scale in Archaeology: Issues of Theory and Practice*. New York: Springer.

Lock, G., Kormann, M., and Pouncett, J. 2014. Visibility and movement: Towards a GIS-based integrated approach. In Polla, S. and Verhagen, P. (eds.), *Computational Approaches to the Study of Movement in Archaeology*. Berlin and Boston: De Gruyter, pp. 23–42. doi:10.1515/9783110288384.23

Lock, G. and Pouncett, J. 2017. Spatial thinking in archaeology: Is GIS the answer?. *Journal of Archaeological Science*, 84, pp. 129–135.

London Charter. 2006. The London Charter for the computer-based visualisation of cultural heritage. http://www.londoncharter.org/

Losier, L.M., Pouliot, J., and Fortin, M. 2007. 3D geometrical modeling of excavation units at the archaeological site of tell 'Acharneh (Syria)'. *Journal of Archaeological Science*, 34(2), pp. 272–288.

Luebke, D.P., Reddy, M., Cohen, J.D., Varshney, A., Watson, B., and Huebner, R. (eds.). 2002. *Level of Detail for 3D Graphics* (Morgan Kaufmann series in computer graphics and geometric modeling). San Francisco: Morgan Kaufmann.

Lundström, V. 2017. *Mapping the Void: A Volumetric Approach to the Vector-Based Line of Sight Analysis of a Pitted Ware Culture Cult House*. Unpublished master's assignment. Lund University.

Madry, S.L.H. and Crumley, C.L. 1990. An application of remote sensing and GIS in a regional archaeological settlement pattern analysis: The Arroux river valley, Burgundy, France. In Allen, K.M.S., Green, S.W., and Zubrow, E.B.W. (eds.), *Interpreting Space: GIS and Archaeology*. London: Taylor and Francis, pp. 364–380.

Margounakis, D. 2008. Virtual reconstructions in archaeology. In Politis, D. (ed.), *E-Learning Methodologies and Computer Applications in Archaeology*. London: IGI Global, pp. 146–156.

Matrone, F., Lingua, A., Pierdicca, R., Malinverni, E., Paolanti, M., Grilli, E., Remondino, F., Murtiyoso, A., and Landes, T. 2020. A benchmark for large-scale heritage point cloud semantic segmentation. *ISPRS – International Archives of the Photogrammetry, Remote Sensing and Spatial Information Sciences*. XLIII-B2-2020. 1419–1426. 10.5194/isprs-archives-XLIII-B2-2020-1419-2020.

McPherron, S.P., Gernat, T., and Hublin, J.-J. 2009. Structured light scanning for high resolution documentation of in situ archaeological finds. *Journal of Archaeological Science*, 36(1), pp. 19–24.

Menache, A. 2000. *Understanding Motion Capture for Computer Animation and Video Games*. Burlington: Morgan Kaufmann.

Mennis, J. 2010. Multidimensional map algebra: Design and implementation of a Spatio-Temporal GIS processing language. *Transactions in GIS*, 14(1), pp. 1–21.

Merlo, S. 2004. Enter the past. The E-way into the four dimensions of cultural heritage. CAA 2003 computer applications and quantitative methods in archaeology. BAR international series 1227. In Ausserer, K. F., Brner, W., Goriany, M. and Karlhuber-Vckl, L. (eds.), *The "Contemporary Mind". 3D GIS as a Challenge in Excavation Practice*. Oxford: Archaeopress, pp. 276–279.

Merlo, S. 2016. *Making Visible: Three-Dimensional GIS in Archaeological Excavation*, BAR International Series 2801 Oxford: Archeopress.

MeshLab: An open-source mesh processing tool. In (V. Scarano, R. De Chiara and U. Erra eds.) Sixth *Eurographics Italian Chapter Conference*. The Eurographics Association, pp. 128–136.

Minozzi, S., Giuffra, V., Bagnoli, J., Paribeni, E., Giustini, D., Caramell, D., and Fornaciari, G. 2010. An investigation of Etruscan cremations by computed tomography (CT). *Antiquity*, 84, pp. 195–201.

Morgan, C.L. and Wright, H.E. 2018. Pencils and pixels: Drawing and digital media in archaeological field recording. *Journal of Field Archaeology*, 43(2), pp. 136–151. doi:10.1080/00934690.2018.1428488.

Moser, J., Albrecht, F., and Kosar, B. 2010. Beyond visualisation -3D GIS analyses for virtual city models. In Kolbe, Th., Konig, G., and Nagel, C. (eds.), *International Archives of the Photogrammetry, Remote Sensing and Spatial Information Sciences XXXVIII-4/W15*, pp. 143–146.

Niccolucci, F. and Hermon, S. 2010. A fuzzy logic approach to reliability in archaeological virtual reconstructions. In Niccolucci, F. and Hermon, S. (eds.), *Beyond the Artefact: Digital Interpretations of the Past (CAA 2004, Proceedings of the 32nd Conference, Prato, Italy, 13–17 April 2004)*, Budapest: Archeolingua, pp. 28–35.

Nobles, G.R. and Roosevelt, C.H. 2021. Filling the void in archaeological excavations: 2D point clouds to 3D volumes. *Open Archaeology*, 7(1), pp. 589–614. doi:10.1515/opar-2020-0149

Nordblad, J. 2012. The shape of history. To give physical form to archaeological Knowledge. In Jensen, O. (ed.), *Histories of Archaeological Practice*. Stockholm: Historiska museet, pp. 241–257.

Ogburn, D.E. 2006. Assessing the level of visibility of cultural objects in past landscapes. *Journal of Archaeological Science*, 33(3), pp. 405–413.

Opitz, R. and Nowlin, J. 2012. Photogrammetric modeling + GIS: Better methods for working with mesh data. http://www.esri.com/news/arcuser/0312/photogrammetric-modeling-plus-gis.html. Accessed 6 February 2021.

Opitz, R. and Cowley, D. (Eds.). 2013. *Interpreting Archaeological Topography: 3D Data, Visualisation and Observation*. Oxford; Oakville: Oxbow Books. doi:10.2307/j.ctvh1dqdz

Opitz, R. and Limp, W.F. 2015. Recent developments in high-density survey and measurement (Hdsm) for archaeology: Implications for practice and theory. *Annual Review of Anthropology*, 44, pp. 347–364. doi:10.1146/annurev-anthro-102214-013845

Opitz, R. and Johnson, T.D. 2016. Interpretation at the controller's edge: Designing Graphical User Interfaces for the digital publication of the excavations at Gabii (Italy). *Open Archaeology*, 2, pp. 1–17.

Opitz, R. 2017. An experiment in using visual attention metrics to think about experience and design choices in past places. *Journal of Archaeological Method and Theory*, 24(4), pp. 1203–1226.

Orengo, H. 2013. Combining terrestrial stereophotogrammetry, DGPS and GIS-based 3D voxel modelling in the volumetric recording of archaeological features. *ISPRS Journal of Photogrammetry and Remote Sensing*, 76, pp. 49–55. doi:10.1016/j.isprsjprs.2012.07.005

Ormsby, T. and Alvi, J., 1999. *Extending ArcView GIS: Teach Yourself to Use ArcView GIS Extensions*. California, USA: Environmental Systems Research Institute Inc., 527 p.

Paliou, E. 2011. The communicative potential of Theran murals in Late Bronze Age Akrotiri: Applying viewshed analysis in 3D townscapes. *Oxford Journal of Archaeology*, 30(3), pp. 247–272.

Paliou, E., Wheatley, D., and Earl, G. 2011. Three-dimensional visibility analysis of architectural spaces: Iconography and visibility of the wall paintings of Xeste 3 (Late Bronze Age Akrotiri). *Journal of Archaeological Science*, 38(2), pp. 375–386.

Paliou, E. and Knight, D.J. 2013. Mapping the senses: Perceptual and social aspects of late antique liturgy in San Vitale, Ravenna. In Contreras Cortés, F., Farjas, M., and Melero, F.J. (eds.), *CAA2010: Fusion of Cultures: Proceedings of the 38th Annual Conference on Computer Applications and Quantitative Methods in Archaeology, Granada, Spain, April 2010*. Oxford: Archaeopress, pp. 229–236.

Paliou, E. 2014. Visibility analysis in 3D built spaces: A new dimension to the understanding of social space. In Paliou, E., Lieberwirth, U., and Polla, S. (eds.), *Spatial Analysis and Social Spaces*. Berlin and Boston: De Gruyter, pp. 91–114.

Papadopoulos, C., Hamilakis, Y., and Kyparissi-Apostolika, N. 2015. Light in a neolithic dwelling: Building 1 at Koutroulou Magoula (Greece). *Antiquity*, 89(347), pp. 1034–1050.

Papadopoulos, C. and Moyes, H. (eds.) 2017. *The Oxford Handbook of Light in Archaeology*. Oxford: Oxford University Press.

Papadopoulos, C., Hamilakis, Y., Kyparissi-Apostolika, N., and Díaz-Guardamino, M. 2019. Digital sensoriality: The Neolithic Figurines from Koutroulou Magoula, Greece. *Cambridge Archaeological Journal*, 29(4), pp. 625–652. doi:10.1017/S0959774319000271

Pavlovskaya, M. 2018. Critical GIS as a tool for social transformation. *The Canadian Geographer/Le Géographe Canadien*, 62(1), pp. 40–54.

Pellini, J.R. 2015. *Coming to Senses: Topics in Sensory Archaeology*. Newcastle upon Tyne: Cambridge Scholars Publishing.

Perry, S., Taylor, J.S., Matsumoto, M., and Uleberg, E. 2018. Theorising the digital: A call to action for the archaeological community. In *CAA2016, Oceans of Data: Proceedings of the 44th Conference on Computer Applications and Quantitative Methods in Archaeology*. Oxford: Archaeopress, pp. 11–22.

Persson, T. 2015. Fornlämning 55:1 Skavet, Inlängan, Torhamn socken, Karlskrona kommun. Arkeologisk undersökning 1989 av skadad hustomtning. *Blekinge Museum Rapport 2015:5.* Karlskrona.

Pigot, S. 1991. Topological models for 3D spatial information systems. In *Technical papers, ACSMASPRS Annual Convention*, Baltimore, Maryland, USA. Auto-Carto 10, 6, pp. 368–392.

Pigot, S. 1992. A topological model for 3D spatial information system, In Bresnahan, P., Corwin, E. and Cowen, D. (ed.), *Spatial Data Handling '92*, Charleston, South Carolina, USA, IGU Commission of GIS, 1, pp. 344–360.

Pilgram, T.K. 1983. *Predicting Archaeological Sites from Environmental Variables: A Mathematical Model for the Sierra Nevada Foothills.* Berkeley: University of California.

Polig, M. 2017. 3D GIS for building archaeology. Combining old and new data in a three dimensional information system in the case study of Lund Cathedral. *Studies in Digital Heritage*, 1(2), pp. 225–238. doi:10.14434/sdh.v1i2.23253

Polig, M., Papacharalambous, D.G., Bakirtzis, N., and Hermon, S. 2020. Assessing visual perception in heritage sites with visual acuity: Case study of the Cathedral of St. John the Theologian in Nicosia, Cyprus.*Journal on Computing and Cultural Heritage (JOCCH)*, 14(1), pp. 1–18.

Powlesland, D. 2016. 3Di – Enhancing the record, extending the returns, 3D imaging from free range photography and its application during excavation. In Kamermans, H., de Neef, W., Piccoli, C., Poluschny, A.G., and Scopigno, R. (eds.), *The Three Dimensions of Archaeology, Proceedings of the XVII UISPP World Congress, Burgos, Spain, 1–7 September 2014.* Oxford: Archaeopress, pp. 13–32.

Primeau, K.E. and Witt, D.E. 2017. Soundscapes in the past: A GIS approach to landscape scale archaeoacoustics. *Archaeology*, 12(2), pp. 245–263.

Raja, R. and Sindbaek, S.M. (eds.). 2018. *Urban Network Evolutions: Towards a High-Definition Archaeology.* Aarhus: ISD LLC.

Raper, J. ed., 1989. *Three Dimensional Applications in GIS.* Boca Raton, Florida, USA: CRC Press.

Re, A., Corsi, J., Demmelbauer, M., Martini, M., Mila, G., and Ricci, C. 2015. X-ray tomography of a soil block: A useful tool for the restoration of archaeological finds. *Heritage Science*, 3(1), pp. 1–7.

Reilly, P. 1989. Data visualization in archaeology. *IBM Systems Journal*, 28(4), pp. 569–579.

Reilly, P. 1991. Towards a virtual archaeology. In Rahtz, S. and K. Lockyear (eds.), *Computer Applications in Archaeology.* Oxford: British Archaeological Reports, pp. 133–139.

Reilly, P. and Rahtz, S.P.Q. (eds.). 1992. *Archaeology and the Information Age: A Global Perspective.* London: Routledge.

Remondino, F. and El-Hakim, S. 2006. Image-based 3D modelling: A review. *The Photogrammetric Record*, 21(115), pp. 269–291.

Renfrew, C. 1997. Foreword. In Forte, M. and Silotti, A. (eds.), *Virtual Archaeology: Re-Creating Ancient Worlds.* New York: Harry Abrams, 7.

Reynoso, C. and Castro, D. 1994. VB-GIS 3D-A GIS specifically designed for archaeology. *Methods in the Mountains, Sydney University Archaeological Methods Series*, 2(2), pp. 135–142.

Rød, J.K. and van der Meer, D. 2009. Visibility and dominance analysis: Assessing a high-rise building project in Trondheim. *Environment and Planning B Planning and Design*, 36(4), pp. 698–710.

Roger, T. 1991. Drowning in data? – Publication and rescue archaeology in the 1990s. *Antiquity*, 65, pp. 822–828. doi:10.1017/S0003598X00080546.

Rood, R.J. 1982. Spatial analysis in archaeology: Historical developments and modern applications. *Lambda Alpha Journal of Man*, 14, pp. 25–60.

Roosevelt, C. Cobb, P., Moss, E., Olson, B., and Ünlüsoy, S. 2015. Excavation is destruction digitization: Advances in archaeological practice. *Journal of Field Archaeology*, 40(3), pp. 325–346. doi:10.1179/2042458215Y.0000000004

Rossi, M., Casali, F., Chirco, P., Morigi, M.P., Nava, E., Querzola, E., and Zanarini, M. 1999. X-ray 3D computed tomography of bronze archaeological samples. *IEEE Transactions on Nuclear Science*, 46, pp. 897–903. doi:10.1109/23.790700

Sacks, R., Eastman, C., Lee, G., and Teicholz, P. 2018. *BIM Handbook: A Guide to Building Information Modeling for Owners, Designers, Engineers, Contractors, and Facility Managers*, 3rd edition. Hoboken: John Wiley & Sons.

Sullivan, E.A., (2020) *Constructing the Sacred Visibility and Ritual Landscape at the Egyptian Necropolis of Saqqara*, Palo Alto, California, USA, Stanford University Press, ISBN 9781503603332. DOI 10.21627/2020cts

Sanders, D.H. 2011. Enabling archaeological hypothesis testing in real time using the REVEAL documentation and display system. *Virtual Archaeology Review*, 2(4), p. 89. doi:10.4995/var.2011.4560.

Scalas, A., Vassallo, V., Mortara, M., Spagnuolo, M., and Hermon, S. 2018. Shape analysis techniques for the Ayia Irini case study. In Sablatnig, R. and Wimmer, M. (eds.), *EUROGRAPHICS: Workshop on Graphics and Cultural Heritage*. (Eurographics Workshop on Graphics and Cultural Heritage). Rome: Eurographics – European Association for Computer Graphics, pp. 255–258. doi:10.2312/gch.20181373

Scharstein, D. and Szeliski, R. 2002. A taxonomy and evaluation of dense two-frame stereo correspondence algorithms. *International Journal of Computer Vision*, 47, pp. 7–42. doi:10.1023/A:1014573219977

Scott, C. B., Roosevelt, C. H., Nobles, Gary R., and Luke, C. 2021. Born-digital logistics: Impacts of 3D recording on archaeological workflow, training, and interpretation. *Open Archaeology*, 7(1), pp. 574–588. https://doi.org/10.1515/opar-2020-0150

Seguchi N. and Dudzik, B. 2019. *3D Data Acquisition for Bioarchaeology, Forensic Anthropology and Archaeology*. Amsterdam: Elsevier.

Seitz, S. M., Curless, B., Diebel, J., Scharstein, D., and Szeliski, R. 2006. A comparison and evaluation of multi-view stereo reconstruction algorithms. In *2006 IEEE Computer Society Conference on Computer Vision and Pattern Recognition (CVPR'06)*, pp. 519–528. doi:10.1109/CVPR.2006.19.

Seville Principles. 2011. The Seville principles, international principles of virtual archaeology. http://sevilleprinciples.com/

Skeates, R. and Day, J. (eds.) 2019. *The Routledge Handbook of Sensory Archaeology*. London: Routledge.

Star, S.L. and Griesemer, J.R. 1989. Institutional ecology, 'Translations' and boundary objects: Amateurs and professionals in Berkeley's museum of vertebrate zoology, 1907–1939. *Social Studies of Science*, 19(3), pp. 387–420.

Staub, T. 2009. Decorative effects and room functions. Evidence of thresholds studied in the residential quarters of insula V 1, Pompeii. *Opuscula*, 2, pp. 205–218.

Streatfeild-James, J. 2016. QGIS for Archaeologists. Getting Started. Guide 42. BAJR Practical Guide Series. http://www.bajr.org/BAJRGuides/42_QGIS_StarterGuide/42_BAJR_Guide_QGIS.pdf

Suleiman W., Joliveau T. and Favier E. 2011. 3D urban visibility analysis with vector GIS data. In: Proceedings of the 19th GIS Research UK Annual Conference, Portsmouth, pp.27–29. Resource document: https://www.researchgate.net/publication/215999859_3D_Urban_Visibility_Analysis_with_Vector_GIS_Data/link/0091ac886cdbd4612adcbea8/download. Accessed 21 June 2021.

Szeliski, R. 2010. *Computer Vision: Algorithms and Applications*. London: Springer-Verlag.

Taylor, J.S. and Issavi, J. 2014. Çatalhöyük research project – Tablet recording workflow: Graphics, *Unpublished Internal Document of the Çatalhöyük Research Project*. Updated 25 June 2014.

Taylor, J.S., Issavi, J., Berggren, Å., Lukas, D., Mazzucato, C., Tung, B., and Dell'Unto, N. 2018. The rise of the machine: The impact of digital tablet recording in the field at Çatalhöyük. *Internet Archaeology*, 47. doi:10.11141/ia.47.1.

Taylor, J. and Dell'Unto, N. 2021. Skeuomorphism in digital archaeological practice: A barrier to progress, or a vital cog in the wheels of change? *Open Archaeology*, 7(1), pp. 482–498. doi:10.1515/opar-2020-0145.

Thomas, J. 1993. The politics of vision and the archaeologies of landscape. In Bender, B. (ed.), *Landscape: Politics and Perspectives* (explorations in anthropology series), pp. 19–48. Routledge.

Thomas, J. 2000. Introduction: The polarities of post-processual archaeology. In Thomas, J. (ed.), *Interpretive Archaeology: A Reader.* New York: Leicester University Press, pp. 1–18.

Thomas, J. 2015. The future of archaeological theory. *Antiquity*, 89(348), pp. 1287–1296.

Tilley, C. 2008. *Body and Image: Explorations in Landscape Phenomenology 2.* Walnut Creek: Left Coast.

Torstensdotter, Å.I. and Söderberg, B. 1994. Skåne, Räng socken, Höllviken 23:7. *RAÄ* 3, 34 och 39. 1992–1992. Arkeologisk utredning och förundersökning. Riksantikvarieämbetet, UV Syd.

Trier, Ø., Larsen, S., and Solberg, R. 2009. Automatic detection of circular structures in high-resolution satellite images of agricultural land. *Archaeological Prospection*, 16(1), pp. 1–15.

Trier, Ø., Zortea, M., and Tonning, C. 2015. Automatic detection of mound structures in airborne laser scanning data. *Journal of Archaeological Science: Reports*, 2, pp. 69–79. ISSN 2352-409X. doi:10.1016/j.jasrep.2015.01.005.

Trier, Ø., Cowley, D., and Waldeland, A. 2018. Using deep neural networks on airborne laser scanning data: Results from a case study of semi-automatic mapping of archaeological topography on Arran, Scotland. *Archaeological Prospection*, 26(2), pp. 165–175. doi:10.1002/arp.1731.

Tringham, R. and Stevanović, M. 2012. *Last House on the Hill*: BACH area reports from Çatalhöyük, turkey. *Monumenta Archaeologica*, 27. Los Angeles: Cotsen Institute of Archaeology (UCLA) Press, pp. 1–594.

Tuan, Y.F. 1977. *Space and Place: The Perspective of Experience.* Minneapolis: University of Minnesota Press.

Valente, R., Brumana, R., Oreni, D., Banfi, F., Barazzetti, L., and Previtali, M. 2017. Object-oriented approach for 3d archaeological documentation. In *26th International CIPA Symposium 2017*, vol. 42. Ottawa: Copernicus Gesellschaft, pp. 707–712.

Van Leusen, M. 1999. *Viewshed and Cost Surface Analysis Using GIS (Cartographic Modelling in a Cell-Based GIS II).* BAR International Series, 757, pp. 215–224.

Van Leusen, M., Deeben, J., Hallewas, D., Zoetbrood, P., Kamermans, H., and Verhagen, P. 2005. A baseline for predictive modelling in the Netherlands. In Van Leusen, M. and Kamermans, H. (eds.), *Predictive Modelling for Archaeological Heritage Management: A Research Agenda.* Amersfoort: NAR, Nederlandse Archeologische Rapporten 29, pp. 25–92.

Van Leusen, M. and Van Gessel, S. 2016. Towards 3D GIS. Notes from the 2012 CAA-NL/DE chapter session "from 2.5 to 3 spatial dimensions. In Kamermans H., de Neef W., Piccoli C., Posluschny A., and Scopigno R. (eds.), *The Three Dimensions of Archaeology. Proceedings of the XVII UISPP World Congress no.* 7. Oxford: Archaeopress Publishing Ltd., pp. 33–37.

Van Riel, S. 2016. Exploring the use of 3D GIS as an analytical tool in archaeological excavation practice. Master's thesis. Lund University. Department of Archaeology & Ancient History. Accessed online 5 April 2021.

Vassallo, V., Landeschi, G., and Hermon, S. (Accepted/In press). Digging in excavation diaries: Digital re-assessment of stratigraphy in 3D GIS. The sanctuary of Ayia Irini, Cyprus. In *Proceedings of the Computer Applications and Quantitative Methods in Archaeology (CAA) international conference, 19–23 March 2018.* Tübingen, Germany.

Verdonck, L., Launaro, A., Vermeulen, F., and Millett, M. 2020. Ground-penetrating radar survey at Falerii Novi: A new approach to the study of Roman cities. *Antiquity*, 94(375), pp. 705–723.

Vergauwen, M. and Van Gool, L. 2006. Web-based 3D reconstruction service. *Machine Vision Applications (MVA)*, 17, pp. 411–426.

Verhagen, P., 2007. Case studies in archaeological predictive modelling (Vol. 14). Leiden University Press.

Verhagen, P. 2018. Spatial analysis in archaeology: Moving into new territories. In Siart C., Forbriger M., Bubenzer O. (eds), *Digital Geoarchaeology.* Cham: Springer, pp. 11–25.

Verhoeven, G. 2011. *Taking Computer Vision Aloft-Archaeological Three-Dimensional Reconstruction from Aerial Photographs with Photoscan.* Wiley Online Library Archaeological prospection 18, pp. 67–73. https://onlinelibrary.wiley.com/doi/full/10.1002/arp.399

Verhoeven, G. 2017. Mesh is more – Using all geometric dimensions for the archaeological analysis and interpretative mapping of 3D surfaces. *Journal of Archaeological Method and Theory*, 24, pp. 999–1033. doi:10.1007/s10816-016-9305-z.

Waagen, J. 2019. New technology and archaeological practice. Improving the primary archaeological recording process in excavation by means of UAS photogrammetry. *Journal of Archaeological Science*, 101, pp. 11–20. ISSN 0305-4403. doi:10.1016/j.jas.2018.10.011

Warren, R.E. 1990. Predictive modelling in archaeology: A primer. In Allen, K.M.S., Green, S.W., and Zubrow, E.B.W. (eds.), *Interpreting Space: GIS and Archaeology.* London: Taylor & Francis, pp. 90–111.

Warren, R.E. and Asch, D.L., 2000. A Predictive Model of Archaeological Site Location in the Eastern Prairie Peninsula. *Practical Applications of GIS for Archaeologists: A Predictive Modeling Kit*, Boca Raton, Florida, United States: CRC Press, pp. 5–32.

Wescott, K.L. and Brandon, R.J. (eds.) 2003. *Practical Applications of GIS for Archaeologists: A Predictive Modelling Toolkit.* London: CRC Press.

Wheatley, D. 1995. Cumulative viewshed analysis: A GIS-based method for investigating intervisibility, and its archaeological application. In Lock, G. and Stancic, Z. (eds.), *Archaeology and GIS: A European Perspective.* London: Taylor and Francis.

Wheatley, D. 2004. Making space for an archaeology of place. *Internet Archaeology*, 15. https://intarch.ac.uk/journal/issue15/wheatley_toc.html

White, D.A. 2016. Archaeology in the age of supercomputing. In: Forte, M. and Campana, S. (eds.), *Digital Methods and Remote Sensing in Archaeology.* Cham: Springer, pp. 323–346.

Wilhelmson, H. and Dell'Unto, N. 2015. Virtual taphonomy: A new method integrating excavation and post-processing of human remains. *American Journal of Physical Anthropology*, 157(2), pp. 305–321. doi:10.1002/ajpa.22715

Witt, D. E. and Primeau, K. E. 2019. March. Performance space, political theater, and audibility in downtown Chaco. In *Acoustics*, vol. 1, no. 1. Basel: Multidisciplinary Digital Publishing Institute, pp. 78–91. https://doi.org/10.3390/acoustics1010007

Witten, A. J., 2017. *Handbook of Geophysics and Archaeology.* United Kingdom: Taylor & Francis Ltd, Sheffield.

Yang, P.P.J., Putra, S.Y., and Li, W. 2007. Viewsphere: A GIS-based 3D visibility analysis for urban design evaluation. *Environment and Planning B: Planning and Design*, 34(6), p. 971.

Zheng, M., Tang, W., Ogundiran, A., Chen, T., and Yang, J. 2020. Parallel landscape visibility analysis: A case study in archaeology. In Tang, W., and Wang, S. (ed.), *High Performance Computing for Geospatial Applications.* Cham: Springer, pp. 77–96.

Zlatanova, S., Rahman, A. and Pilouk, M., 2002. 3D GIS: current status and perspectives. In (edited by C. Armenakis, Y.C. Lee) International Archives of Photogrammetry Remote Sensing and Spatial Information Sciences, 34(4), pp.66-71. Resource document. https://www.isprs.org/proceedings/XXXIV/part4/. Accessed 22 June 2021.

Zubrow, E.B.W. 2010. From Archaeology to I-archaeology: Cyberarchaeology, paradigms, and the end of the twentieth century. In Forte, M. (ed.), *Cyber Archaeology.* Oxford: Archaeopress BAR International Series 2177, pp. 1–8.

Index

Note: **Bold** page numbers refer to tables and *italic* page numbers refer to figures.